# Advanced Master Keying Skills

**by Don O'Shall and John Truempy**
based on research by Greg Brockway

Locksmithing Education Publishing, LLC
Second Edition, August 2017
**(First Edition, © January 2006)**
by Donald O'Shall and John Truempy,
Beverly Hills, FL - USA



**ISBN 978-1-937067-23-6**

**Library of Congress Cataloguing-in-Publication Data**

O'Shall, Donald and
Truempy, John

Advanced Master Keying Skills / by Don O'Shall and John Truempy

**ISBN 978-1-937067-23-6**

1. Industry—Security Measures—Management

ISBN 9781937067236 **$65.00 USD**
56500 >
9 781937 067236

**Locksmithing Education Publishing, LLC**
12 South Lee Street
Beverly Hills, FL - 34465

# Advanced Masterkeying Textbook

# INTRODUCTION

**If you are reading this, you should already have an understanding of Basic Masterkeying techniques, as well as a firm understanding of the "Base Four" ( Two Increment Drop ) and "Base Five" ( Single Increment Drop ) charts and the Incidental Masterkeys shown on them. You should already understand the Sequence of Progression and the rules applied to each position of the progression based on them. You should understand the Key Bitting Array and the effect that varying the sequence within the KBA has on when the bittings appear in the charts.**

**You should understand the Standard Key Coding System (SKCS) and how and when to apply it to the bittings, cylinders and keys you develop. You should understand the concept of Blind Coding for CKC (Concealed Key Coding) and know at least a few techniques to develop or implement a Blind Code.**

**You should understand the advantages and disadvantages of various type of key stamping and key tracing and the importance of a TMK Log. You should know the proper procedures for storing customer information to provide a reasonable level of privacy for their keying information.**

**You should understand the importance of proper key issue and return procedures and the key control document, which you should work with your client/customer to develop and implement.**

**If any of that is not fully familiar to you, it is VITAL that you bring yourself up to speed with that prior to reading this book or taking the courses associated with it.** To proceed further without the basics solidly under your belt would be to possibly confuse yourself to the point where your systems will have the potential to put your customers and your reputation as a professional at risk.

That having been said, if you are still reading, we will assume that you do understand those basics fully and are ready to proceed with advanced masterkeying techniques.

# A word about the terminology in this book.

In the **Master Keying Text Book Second Edition**, we were able to use industry approved terminology and the definitions from the LIST (Lock Industry Standards and Training) Council almost entirely throughout the book, with the exception being terms that were unique to the teaching style used by the book.

That will not be the case in this book, however, because most of the techniques in this book have no industry approved terminology at the time of this writing. This is because for a term to be approved for the LIST Council dictionary, it must be a) accurate and b) in common usage in the industry. Too few people in the industry are currently aware of the techniques shown in this book to consider any terminology related to them to meet section b of the requirements.

Billy B. Edwards, John Truempy and myself are on the LIST Council and I am certain we would ALL like to see standardized terminology relating to Advanced Master Keying. But it is not our position to create new terminology where none exists and stuff it down the throats of locksmiths. The Council has never done that, and I hope never will.

As more locksmiths become aware of the techniques of Advanced Master Keying, the terminology will adapt itself to their needs, and the LIST Council will adopt appropriate terminology. At that time, some of the terminology in this book may become outdated and it will become necessary to re-write this book with the corrected terminology. Until then, I have attempted to use the more popular terms whenever possible, and to list additional terms of which I am aware.

In this book, we touch gently upon the technique known as Matrix. Although the general technique can be traced back to Vic Fanberg in the late 1970's, and the topic is shown in virtually all advanced master keying related books, articles and courses including those offered by manufacturers, the indisputable father of the technique as it is used today is Billy B. Edwards, and the reader is directed toward his latest book on Master Keying to really learn the technique and its full potential. At the time of the writing of this book, his latest book is:

# Prelude

**Before we move ahead we should take a moment to remember all the Masterkeying Experts who brought us to this point.** The list is far too long to put here in its entirety, of course, so I will merely list some of those who directly influenced my own development. The following is excerpted from a speech I made in 1999 on the topic of Masterkeying.

Masterkeying is a well established process. According to **Tom Hennessy**, (curator of the **Lock Museum of America** in Terryville, Connecticut since its foundation in 1972 and a renowned masterkeying expert), Masterkeying of Lever Locks can be traced back to as early as 1686 , and Masterkeying of Pin Tumbler Locks appeared in an 1865 Yale & Greenleaf advertisement.

But in spite of that, it was not until the 1950's when mainstream locksmiths first began to seriously evaluate its capabilities and limitations and the methods then in use to accomplish it. The **Locksmith Ledger** published its first masterkeying article using the techniques on which modern masterkeying is based, and serious discussion and investigation of the topic began. **Jerome Brush** put out the **Ilco Master Key Selector** to lay out small systems.

And it was nearly twenty years later before innovative locksmiths (such as **Wiegend Jensen, Ken Ehrenreich, Bert Michaels, Hollis DeVines, Gerry Finch, Cal Brittel, and others too numerous to mention here**, but whose efforts were more than notable) began to explore new methods of creating the systems, and to explore questions pertaining to those perceived capabilities and limitations, and more importantly, to share their observations with others, both in person and in print (in magazines such as **The National Locksmith** and the **Locksmith Ledger**, as well as in books and courses designed primarily for the mainstream Locksmiths).

## Knowledge is built upon knowledge. Honor the past and you prepare for the future.

***This is dedicated, like a drop of fine wine upon the ground, to those who have gone before us, without whom our progress would not be possible.***

# The Grand Masters Of Master Keying

Lenny Singer Tom Hennessy Harold George John Wiedman Hollis De Vines

Wiegend Jensen Gerry Connelly Ken Ehrenreich Gerry Finch Jerome Andrews

Billy Edwards A.J. Hoffman James Watt Don O'Shall John Truempy

NOT SHOWN: Cal Brittel, Bert Michaels, Tony West

Inspired by these men, (The above list represents those who directly influenced my own education and direction.) a new generation of Masterkeying Specialists emerged, and began to organize methods, step options, formats and styles that maximized the usefulness of Professional Masterkeying Techniques to the public at large, as well as the important task of sharing that information not only with the mainstream Locksmith, but with a new generation of Institutional Locksmiths as well.

Once again, the list of people who should be included here is too numerous to mention, so I will concentrate on those with whom I had direct and personal dealings, often challenging each other to develop ever-increasing levels of proficiency, in many cases answering questions that had not been even asked before. These included **Billy B. Edwards, Anthony West, A. J. Hoffman, Jerome V. Andrews, Edward M. Greene, Breck Camp, Mike Kirkpatrick, Gale Johnson, Greg Brockway, John Truempey, Lloyd Seliber, and of course, myself.**

I have every faith that our works will inspire yet another generation of research on the topic, and add new names to the list of those who influenced the professional techniques of Masterkeying.

-Don O'Shall, in a speech made in March 1999.

That faith has by the time of this writing, inspired two other locksmiths to take up the mantle. Eric Costley, younger brother of renowned lock and safe instructor Brian Costley, has been researching and writing on the Spherical Method of Master Keying. Rod Oden has been writing on the effect of the effect of Generics and Associated Master Keys on the Spherical Method. It is this author's fervent and constant hope that others will become inspired by these writings and the work that preceded them, and we will have a new generation of Master Keying experts to write and teach for the good of the Locksmithing profession.

- Don O'Shall, from a speech made in 2005.

Noted Master Key expert Billy B. Edwards Jr.
(http://www.thelockman.com/) on December 01, 2005 stated:

"The teaching of Master Keying has gone through some changes over the years. Back in the late 1960's it was being taught by people like Cal Brittel, Hollis Devines, Harold George and Weigand Jensen. In those days Total Position Progression was the only method taught and I would venture to say that was because the Rotating Constant Method was unknown outside of manufacturing.

In fact, we determined from very old Yale catalogs that the first master key systems, (those offered by Yale up until 1872), were in fact shoebox key systems. That was evident by the catalog claim that an unlimited number of Keyed Different locks could also be operated by a single Master Key.

Upon the discovery of interchange in these systems, Yale discontinued Master Keying from the catalog and invented the Duplex lock. The duplex lock was a mortise lock body that had two mortise cylinders that could operate the same bolt. Their answer to the need for a Master Key, one cylinder was Keyed Different while the other was Keyed Alike for the entire site. They did not develop a Triplex lock for Grand Master Keying.

By 1878, Yale keying personnel had discovered the principles of the Rotating Constant Method and routinely implemented four constants in their five pin cylinders. That fact was evident by the catalog claim that you could have 1 Master Key with up to 20 Change Keys.

There is in fact an engineering drawing labeled #4557E and 4457E dated 9/11/1896 that indicates they discovered the Total Position Progression that year and that the Total Position Progression was used for the Surety System.

Once the predecessor of the LIST Council started its work toward making a Master Keying class we discovered that we first needed a standardized glossary of terms. Weigand Jensen was a charter member of that group and I think it was at a meeting in Dallas that we all had the epiphany that the only two methods of progression were Total Position Progression and Rotating Constant Method.

During those times Master Keying was being taught by Ken Ehrenreich nationally and he happened to also be the Chairman of the group, and he only taught Total Position Progression. That group was the first to recommend that not every system had to have the full 1024 or 4096 compliment of Change Keys and that you should in fact use as few progressives as possible to produce the desired system plus about 50-100% expansion.

Ken Ehrenreich adopted that method in his class and later it was furthered by his assistant Hal Tibbets when he took over teaching the class. We also included that technique of limiting the progression in our certified class, (no Incidental Master Keys to avoid confusing the students), and also introduced them to Rotating Constant Method via the two limited progression pages we had them make in class.

Those two pages are revealed in the class to be just two of the twenty patterns of progression available for a 3 in 6 Rotating Constant Method of progression. In both cases the student realizes that this system only has 2 levels of keying and he is specifically told that a three level system isn't possible with the patterns of progression utilized. He is also told to never use the Rotating Constant Method beyond 2 levels of keying until he has had a more advanced class.

We also indicate that they can use other Rotating Constant Method options to do 3 levels of keying if they divide the key and use Rotating Constant Method for only part of the key, i.e. 1 in 3, 2 in 3, 1 in 4, etc. the important thing is that they eventually realize that the Rotating Constant Method is using some of the 11,528 Master Keys from a Total Position Progression system and not the Change Keys from a Total Position Progression system, but that is another class."
BBE (Billy B. Edwards)

# Chapter One: Parallel Master Keys

On our standard (SPF) charts we can easily see the **Block Master Keys**. They appear just below the block they operate, and are created by using the cut from the TMK ( Top Master Key for the entire charted system ) from progression sequence **a ( the first progression sequence position ).** They are quite easy to calculate and useful in many systems.

| | | | | | | |
|---|---|---|---|---|---|---|
| 1 | 5 | 4 | 3 | 2 | 7 | 0 |
| 2 | 5 | 4 | 3 | 2 | 7 | 2 |
| 3 | 5 | 4 | 3 | 2 | 7 | 4 |
| 4 | 5 | 4 | 3 | 2 | 7 | 6 |

BLOCK MASTER

To calculate them, we simply use the cut from the TMK (Top or TMK, depending on the technique being used) from position a (the FIRST Progression Sequence) and combine it with all the other cut depths the bittings in the block have in common.

In the example above, we see that the position that changes with every new bitting in the block is on the far right, indicating that position is the first progression sequence. It uses a 0, 2, 4, and 6 so obviously the TMK cut for that position is the remaining even cut possibility of 8. The change keys in this block all have 5 4 3 2 and 7, so the Block Master is 5 4 3 2 7 8 for this block.

It is very simple and very easy to see on the chart.

On most standard charts, we can also easily see the **Line Master Keys**. They appear just to the right of the line or row they operate, and are created by using the cut from the TMK ( Top Master Key for the entire charted system ) from progression sequence **c ( the third progression sequence position )** with the cuts the change keys involved have in common. They are also quite easy to calculate and useful in many systems.

| | | | | | | | | | | | | | | | | | | | | | | | | | | | |
|---|---|---|---|---|---|---|---|---|---|---|---|---|---|---|---|---|---|---|---|---|---|---|---|---|---|---|---|
| 1 | 5 | 4 | 3 | 2 | 7 | 0 | 17 | 5 | 4 | 3 | 4 | 7 | 0 | 33 | 5 | 4 | 3 | 6 | 7 | 0 | 49 | 5 | 4 | 3 | 8 | 7 | 0 |
| [illegible] | [illegible] | [illegible] | [illegible] | [illegible] | [illegible] | [illegible] | [illegible] | [illegible] | [illegible] | [illegible] | [illegible] | [illegible] | [illegible] | [illegible] | [illegible] | [illegible] | [illegible] | [illegible] | [illegible] | [illegible] | [illegible] | [illegible] | [illegible] | [illegible] | [illegible] | [illegible] | [illegible] |

The Line Master above would be 5 4 3 0 7 0. The Line Master (LMK) has been referred to in some texts as the row master, but this often gets confused with the Horizontal Master Key which other texts have called a Horizontal Row Master. Therefore the Line Master Key is the commonly preferred title for this level of master key.

There is a master key for each Vertical Column of change key bittings that typically appears on the SPF charts. It uses the TMK cuts from positions a and b (first and second progression sequences) with the remaining cuts the bittings in that column have in common.

| | VERTICAL MASTER | | | | | |
|---|---|---|---|---|---|---|
| 1 | 5 | 4 | 3 | 2 | 7 | 0 |
| 2 | 5 | 4 | 3 | 2 | 7 | 2 |
| 3 | 5 | 4 | 3 | 2 | 7 | 4 |
| 4 | 5 | 4 | 3 | 2 | 7 | 6 |
| | BLOCK MASTER | | | | | |
| 5 | 5 | 4 | 3 | 2 | 9 | 0 |
| 6 | 5 | 4 | 3 | 2 | 9 | 2 |
| 7 | 5 | 4 | 3 | 2 | 9 | 4 |
| 8 | 5 | 4 | 3 | 2 | 9 | 6 |
| | BLOCK MASTER | | | | | |
| 9 | 5 | 4 | 3 | 2 | 1 | 0 |
| 10 | 5 | 4 | 3 | 2 | 1 | 2 |
| 11 | 5 | 4 | 3 | 2 | 1 | 4 |
| 12 | 5 | 4 | 3 | 2 | 1 | 6 |
| | BLOCK MASTER | | | | | |
| 13 | 5 | 4 | 3 | 2 | 3 | 0 |
| 14 | 5 | 4 | 3 | 2 | 3 | 2 |
| 15 | 5 | 4 | 3 | 2 | 3 | 4 |
| 16 | 5 | 4 | 3 | 2 | 3 | 6 |
| | BLOCK MASTER | | | | | |

The VMK for the above chart would be 5 4 3 2 5 8.

There is a similar level of master key for the four blocks across the chart which make up a Horizontal Column. It uses the TMK cuts from sequence positions a and c (first and third progression sequences) with the remaining change key cuts that the bittings in that horizontal column have in common.

543078
HORIZONTAL MASTER

| | | | | | | | | | | | | | | | | | | | | | | | | | | |
|---|---|---|---|---|---|---|---|---|---|---|---|---|---|---|---|---|---|---|---|---|---|---|---|---|---|---|
| 1 | 5 | 4 | 3 | 2 | 7 | 0 | 17 | 5 | 4 | 3 | 2 | 7 | 2 | 33 | 5 | 4 | 3 | 2 | 7 | 4 | 49 | 5 | 4 | 3 | 2 | 7 | 6 |
| 2 | 5 | 4 | 3 | 4 | 7 | 0 | 18 | 5 | 4 | 3 | 4 | 7 | 2 | 34 | 5 | 4 | 3 | 4 | 7 | 4 | 50 | 5 | 4 | 3 | 4 | 7 | 6 |
| 3 | 5 | 4 | 3 | 6 | 7 | 0 | 19 | 5 | 4 | 3 | 6 | 7 | 2 | 35 | 5 | 4 | 3 | 6 | 7 | 4 | 51 | 5 | 4 | 3 | 6 | 7 | 6 |
| 4 | 5 | 4 | 3 | 8 | 7 | 0 | 20 | 5 | 4 | 3 | 8 | 7 | 2 | 36 | 5 | 4 | 3 | 8 | 7 | 4 | 52 | 5 | 4 | 3 | 8 | 7 | 6 |

5 4 3 0 7 0    5 4 3 0 7 2    5 4 3 0 7 4    [illegible]

Note that the HMK for this horizontal column is shown as 5 4 3 0 7 8

The next level of master key shown on the chart is the Page Master, which uses the three TMK cuts from the first three progression sequences ( a, b and c) with the cuts the rest of the change key bittings on a page of the chart have in common.

| | | | | | | |
|---|---|---|---|---|---|---|
| 3 | 2 | 1 | 0 | 5 | 8 | THEORETICAL MASTER |
| 5 | 4 | 3 | 2 | 7 | 0 | BASE FIELD (KBA) KEY BITTING ARRAY |
| 7 | 6 | 5 | 4 | 9 | 2 | |
| 9 | 8 | 7 | 6 | 1 | 4 | |
| 1 | 0 | 9 | 8 | 3 | 6 | |
| f | e | d | a | b | c | ⇦ SEQUENCE OF PROGRESSION |

PAGE MASTER 54 3 058 — PAGE # 1 OF 64 PAGES
GROUP MASTER 54 1058
SECTION MASTER 5 21058

| | # | 54 3 050 VERTICAL MASTER | # | 54 3 052 VERTICAL MASTER | # | 54 3 054 VERTICAL MASTER | # | 54 3 056 VERTICAL MASTER | LINE MASTERS ⇩ |
|---|---|---|---|---|---|---|---|---|---|
| | 1 | 543270 | 17 | 543272 | 33 | 543274 | 49 | 543276 | 543 278 |
| 543078 HORIZONTAL MASTER | 2 | 543470 | 18 | 543472 | 34 | 543474 | 50 | 543476 | 54 3 478 |
| | 3 | 543670 | 19 | 543672 | 35 | 543674 | 51 | 543676 | 54 3678 |
| | 4 | 543870 | 20 | 543872 | 36 | 543874 | 52 | 543876 | 54 3878 |
| | | 54 3 0 7 0 BLOCK MASTER | | 54 3 0 7 2 BLOCK MASTER | | 54 3 0 7 4 BLOCK MASTER | | 54 3 0 7 6 BLOCK MASTER | |
| | 5 | 543290 | 21 | 543292 | 37 | 543294 | 53 | 543296 | 54 3298 |
| 543098 HORIZONTAL MASTER | 6 | 543490 | 22 | 543492 | 38 | 543494 | 54 | 543496 | 54 3498 |
| | 7 | 543690 | 23 | 543692 | 39 | 543694 | 55 | 543696 | 54 3698 |
| | 8 | 543890 | 24 | 543892 | 40 | 543894 | 56 | 543896 | 54 3898 |
| | | 54 3 0 9 0 BLOCK MASTER | | 54 3 0 9 2 BLOCK MASTER | | 54 3 0 9 4 BLOCK MASTER | | 54 3 0 9 6 BLOCK MASTER | |
| | 9 | 543210 | 25 | 543212 | 41 | 543214 | 57 | 543216 | 54 3218 |
| 543018 HORIZONTAL MASTER | 10 | 543410 | 26 | 543412 | 42 | 543414 | 58 | 543416 | 54 3418 |
| | 11 | 543610 | 27 | 543612 | 43 | 543614 | 59 | 543616 | 54 3618 |
| | 12 | 543810 | 28 | 543812 | 44 | 543814 | 60 | 543816 | 54 3818 |
| | | 54 30 1 0 BLOCK MASTER | | 54 3 0 1 2 BLOCK MASTER | | 54 3 0 1 4 BLOCK MASTER | | 54 3 0 1 6 BLOCK MASTER | |
| | 13 | 543230 | 29 | 543232 | 45 | 543234 | 61 | 543236 | 54 3238 |
| 543038 HORIZONTAL MASTER | 14 | 543430 | 30 | 543432 | 46 | 543434 | 62 | 543436 | 54 3438 |
| | 15 | 543630 | 31 | 543632 | 47 | 543634 | 63 | 543636 | 54 3638 |
| | 16 | 543830 | 32 | 543832 | 48 | 543834 | 64 | 543836 | 54 3838 |
| | | 54 3 0 3 0 BLOCK MASTER | | 54 3 0 3 2 BLOCK MASTER | | 54 3 0 3 4 BLOCK MASTER | | 54 3 0 3 6 BLOCK MASTER | |

Note the Page Master on the chart above is 5 4 3 <u>0</u> <u>5</u> <u>8</u>. This is true even though the Sequence of Progression is not the popular right to left pattern, but begins in the middle and goes to the right, then goes back to the middle and works again to the left. The three on the right in this example are still the first, second and third sequences (a, b and c).

Typically above that level are shown the group master key which uses the TMK cuts from progression sequences a, b, c and d (the first four progression sequences) and the section master key, which uses the TMK cuts from progression sequences a, b, c, d and e (the first five progression sequences) with the remaining position cut(s).

These are usually the only master keys shown on the SPF charts.

These master keys are **Incidental Master Keys**. They appear in the system and work automatically if the cylinders are pinned to the change key and the Top (or TMK). They are easy to see on the chart, and the area they are capable of operating is clearly defined.

But there are also other Incidental Master Keys of the same levels that are not as easy to see or to put logically on the charts. We created the Block and Line Master Keys by using the master key cut from one position of the TMK, from positions **a ( first )** and **c ( third)** respectively. But there are also positions **b ( second ), d ( fourth ), e (fifth ), and f (sixth),** as well as **position g ( the seventh )** on a seven pin cylinder.

These Incidental Master Keys which do not appear on the chart, but which could be created in order to satisfy specific needs within the system were once called **Ghost Keys** or **Phantom Keys**, as were the other Incidental Master Keys which do appear on the charts. However, this was not an accurate representation of their role in the system, and those terms were abandoned.

For many years, the incidental master keys which do not appear on the chart remained un-named. They were finally named by Jerome V. Andrews, who coined the term, **PARALLEL MASTER KEYS.** The term was based on the fact that they are incidental master keys which are created parallel to the development of the chart, and thus do not appear on it.

It should be noted that at the time of this writing the term Parallel Master Keys has not been approved and added to the LIST Council Dictionary. Their logical stance thus far is that they are incidental master keys whether they appear on a chart or not.

**Because they do not appear on the chart, it is not easy to give them a descriptive term such as Block Master or Line Master. For this reason they are usually referred to by the position(s) they use from the TMK.** A Parallel Master Key, for example, that uses cuts from the second ( b ) and third (c ) positions would be considered a **bc Parallel Master Key.**

Looking at a standard SPF chart, it is easy to see the four bitting combinations operated by the Block Master (a) beneath the block, or by the Line Master (c), but the b master will also control four bitting combinations, by using the TMK cut from position b with the cuts the other positions have in common.

Therefore, we can see that the first one will be 5 4 3 2 5 0, and that it will operate the bitting combination identified on the chart as "1", the one identified as "5", the one identified as "9", and the one identified as "13", or the bittings:

5 4 3 2 7 0
5 4 3 2 9 0
5 4 3 2 1 0
5 4 3 2 3 0

| 3 | 2 | 1 | 0 | 5 | 8 | THEORETICAL MASTER |
|---|---|---|---|---|---|---|
| 5 | 4 | 3 | 2 | 7 | 0 | BASE FIELD |
| 7 | 6 | 5 | 4 | 9 | 2 | (KBA) |
| 9 | 8 | 7 | 6 | 1 | 4 | KEY BITTING ARRAY |
| 1 | 0 | 9 | 8 | 3 | 6 | |
| f | e | d | c | b | a | ⇦ SEQUENCE OF PROGRESSION |

PAGE MASTER ______
GROUP MASTER ______
SECTION MASTER ______

PAGE # 1 OF 64 PAGES

| | VERTICAL MASTER | | VERTICAL MASTER | | VERTICAL MASTER | | VERTICAL MASTER | | LINE MASTERS ⇓ |
|---|---|---|---|---|---|---|---|---|---|
| 1 | 5 4 3 2 7 0 | 17 | 5 4 3 4 7 0 | 33 | 5 4 3 6 7 0 | 49 | 5 4 3 8 7 0 | | |
| 2 | 5 4 3 2 7 2 | 18 | 5 4 3 4 7 2 | 34 | 5 4 3 6 7 2 | 50 | 5 4 3 8 7 2 | | |
| HORIZONTAL MASTER 3 | 5 4 3 2 7 4 | 19 | 5 4 3 4 7 4 | 35 | 5 4 3 6 7 4 | 51 | 5 4 3 8 7 4 | | |
| 4 | 5 4 3 2 7 6 | 20 | 5 4 3 4 7 6 | 36 | 5 4 3 6 7 6 | 52 | 5 4 3 8 7 6 | | |
| | BLOCK MASTER | | BLOCK MASTER | | BLOCK MASTER | | BLOCK MASTER | | |
| 5 | 5 4 3 2 9 0 | 21 | 5 4 3 4 9 0 | 37 | 5 4 3 6 9 0 | 53 | 5 4 3 8 9 0 | | |
| 6 | 5 4 3 2 9 2 | 22 | 5 4 3 4 9 2 | 38 | 5 4 3 6 9 2 | 54 | 5 4 3 8 9 2 | | |
| HORIZONTAL MASTER 7 | 5 4 3 2 9 4 | 23 | 5 4 3 4 9 4 | 39 | 5 4 3 6 9 4 | 55 | 5 4 3 8 9 4 | | |
| 8 | 5 4 3 2 9 6 | 24 | 5 4 3 4 9 6 | 40 | 5 4 3 6 9 6 | 56 | 5 4 3 8 9 6 | | |
| | BLOCK MASTER | | BLOCK MASTER | | BLOCK MASTER | | BLOCK MASTER | | |
| 9 | 5 4 3 2 1 0 | 25 | 5 4 3 4 1 0 | 41 | 5 4 3 6 1 0 | 57 | 5 4 3 8 1 0 | | |
| 10 | 5 4 3 2 1 2 | 26 | 5 4 3 4 1 2 | 42 | 5 4 3 6 1 2 | 58 | 5 4 3 8 1 2 | | |
| HORIZONTAL MASTER 11 | 5 4 3 2 1 4 | 27 | 5 4 3 4 1 4 | 43 | 5 4 3 6 1 4 | 59 | 5 4 3 8 1 4 | | |
| 12 | 5 4 3 2 1 6 | 28 | 5 4 3 4 1 6 | 44 | 5 4 3 6 1 6 | 60 | 5 4 3 8 1 6 | | |
| | BLOCK MASTER | | BLOCK MASTER | | BLOCK MASTER | | BLOCK MASTER | | |
| 13 | 5 4 3 2 3 0 | 29 | 5 4 3 4 3 0 | 45 | 5 4 3 6 3 0 | 61 | 5 4 3 8 3 0 | | |
| 14 | 5 4 3 2 3 2 | 30 | 5 4 3 4 3 2 | 46 | 5 4 3 6 3 2 | 62 | 5 4 3 8 3 2 | | |
| HORIZONTAL MASTER 15 | 5 4 3 2 3 4 | 31 | 5 4 3 4 3 4 | 47 | 5 4 3 6 3 4 | 63 | 5 4 3 8 3 4 | | |
| 16 | 5 4 3 2 3 6 | 32 | 5 4 3 4 3 6 | 48 | 5 4 3 6 3 6 | 64 | 5 4 3 8 3 6 | | |
| | BLOCK MASTER | | BLOCK MASTER | | BLOCK MASTER | | BLOCK MASTER | | |

The second b parallel master key (5 4 3 2 5 2) would be for the second bittings in each of the four blocks:

5 4 3 2 7 2
5 4 3 2 9 2
5 4 3 2 1 2
5 4 3 2 3 2

There will be four b parallel master keys under the first vertical column master, and 16 on the page, under the page master key. Each operates one bitting combination per block in a vertical column.

This simple example shows the rules for handling of ALL incidental master keys, whether they appear on the chart or are parallel to it.

## 1. Comparing the cuts of the bittings you wish the master to be capable of operating, wherever they have the same cuts, the incidental master key will use those cuts, and wherever any differences occur, it will use the TMK cut for that position.

## 2. They will be logical in size.

If they use one cut from the TMK, they will operate as many change key bittings as the KBA (Key Bitting Array) has in that position, not including the TMK cut. If it is a ten depth two increment drop (standard two step), this will be four. If it is a 1 to 6 or 0 to 5 single increment drop, this will be five.

If they use two cuts from the TMK, they will operate as many change key bittings as the KBA (Key Bitting Array) has in that position, not including the TMK cut multiplied by the quantity from the KBA in the previous progression sequence. If it is a ten depth two increment drop (standard two step), this will be 16. If it is a 1 to 6 or 0 to 5 single increment drop, this will be 25.

If they use three cuts from the TMK, they will operate as many change key bittings as the KBA (Key Bitting Array) has in that position, not including the TMK cut, multiplied by the quantities from the KBA in two previous progression sequences. If it is a ten depth two increment drop (standard two step), this will be 64. If it is a 1 to 6 or 0 to 5 single increment drop, this will be 125.

If they use four cuts from the TMK, they will operate as many change key bittings as the KBA (Key Bitting Array) has in that position, not including the TMK cut, multiplied by the quantities from the three previous KBA positions. If it is a ten depth two increment drop (standard two step), this will be 256. If it is a 1 to 6 or 0 to 5 single increment drop, this will be 625.

If they use five cuts from the TMK, they will operate as many change key bittings as the KBA (Key Bitting Array) has in that position, not including the TMK cut, multiplied by the quantities from the KBA in the previous four positions. If it is a ten depth two increment drop (standard two step), this will be 1,024. If it is a 1 to 6 or 0 to 5 single increment drop, this will be 3,125.

## 3. They will be logical in position.

An incidental master key (whether parallel or not) will have a relational position. If it operates a "first" bitting, every bitting it operates will be a "first" e.g. first combination per block, or first block per column, or first column per page, etc.

For example, the first d parallel master key in a two increment drop with ten depth possibilities operates the first bitting in the first block per page for the first four pages of the first group in the first section of a six pin system.

The second d parallel master key in a two increment drop with ten depth possibilities operates the second bitting from the first block per page for the first four pages of the first group in the first section of a six pin system.

And so it continues. But the easiest way to look at it is ALWAYS to simply compare the bittings to each other and use the TMK cut from the positions where the differences occur!

Interestingly enough, the more commonly usable parallel master keys tend to be the more logical ones. Although there are indeed applications for a bdef parallel key, for example, which might operate the top combination from every block in the first column of every page in the system (more about that later), it is much more common to find a use for an adef parallel master key, which operates the top block of every page in the system.

For example in a master key for several athletic buildings, there is frequently a need for a selective master key to be issued that can handle one or two rooms in each building for the Sports Medicine people.

Because this is crossing master keys, it could be done with a severe reduction in physical security within the cylinders using standard industry practices for cross-keying.

But by issuing the adef key, you simply pull the bittings from the first block of every affected page when the Sports Medicine Master Key operates it, and be careful NOT to pull it from that block if it does not. There is no reduction in physical integrity within the cylinder and no loss of security involved beyond that of a normal master keyed cylinder in the system.

There are multiples of almost every parallel master key, just as there were multiples of the Block Master Key and multiples of the Line Master Key appearing on the SPF chart. But as we saw with those, there were more of the lower level masters such as the Block Master Keys, than there were of the higher level masterkeys such as the Page Master Key.

If it were necessary, we could describe most, such as the bdef parallel master we used as an example earlier, which operated the top combination of every block in the first column of every page. It has a similar bdef parallel master for the second position of every block in the first column of every page. And there is another for the third position of every block in the first column of every page. Also, there is a bdef parallel master key for the fourth position of every block in the first column of every page.

Now here comes the big news. There are another four just like the first four, except that they are one combination in each block of the second column of every page. And there are another four for the four positions in every block of the third column on every page, and finally, another four for the fourth column.

That represents all the possible bdef parallel master keys, a total of 16 on a two increment drop six pin, or with five positions in every block of a single increment drop, there would be 25 bdef parallel master keys.

Those numbers are important, because they make the identification of what each parallel master key does very simple and obvious. Instead of paying attention to the positions where the parallel master key uses cuts from the TMK, concentrate instead on the position(s) where it does not, because these are the positions that define its scope.

That means that our bdef key from earlier is defined in its scope by the first and third positions (a and c respectively). Using a six pin key, the bdef parallel master key would operate every combination that has the same cut numbers in positions a and c as does the bdef parallel master key.

As an example, a TMK of 321058 using a two increment drop, and right to left progression sequence with an odometer arranged Key Bitting Array, would yield four bdef parallel master keys: 321250, 321252, 321254 and 321256 respectively, for the first column.

The second column would yield similar bdef parallel master keys of 321450, 321452, 321455, and 321456 respectively.

Each of the eight we have seen so far operates everything that matches its particular cuts in positions a and c (first and third progression sequences). Therefore, the 321250 would operate the first combination in every block in the first column of every page, because those are the combinations that have a 2 in the third progression sequence and a 0 in the first progression sequence.

Similarly, the 321456 would operate the fourth combination in every block in the second column on every page, because those are the combinations that have a 4 in the third progression sequence and a 6 in the first.

In order to see this more clearly, let's look at the bittings on that page:

| VM | 5 4 3 2 5 8 | VM | 5 4 3 4 5 8 | VM | 5 4 3 6 5 8 | VM | 5 4 3 8 5 8 |
|---|---|---|---|---|---|---|---|
| | 5 4 3 2 7 0 | | 5 4 3 4 7 0 | | 5 4 3 6 7 0 | | 5 4 3 8 7 0 |
| | 5 4 3 2 7 2 | | 5 4 3 4 7 2 | | 5 4 3 6 7 2 | | 5 4 3 8 7 2 |
| | 5 4 3 2 7 4 | | 5 4 3 4 7 4 | | 5 4 3 6 7 4 | | 5 4 3 8 7 4 |
| | 5 4 3 2 7 6 | | 5 4 3 4 7 6 | | 5 4 3 6 7 6 | | 5 4 3 8 7 6 |
| BM | 5 4 3 2 7 8 | BM | 5 4 3 4 7 8 | BM | 5 4 3 6 7 8 | BM | 5 4 3 8 7 8 |
| | 5 4 3 2 9 0 | | 5 4 3 4 9 0 | | 5 4 3 6 9 0 | | 5 4 3 8 9 0 |
| | 5 4 3 2 9 2 | | 5 4 3 4 9 2 | | 5 4 3 6 9 2 | | 5 4 3 8 9 2 |
| | 5 4 3 2 9 4 | | 5 4 3 4 9 4 | | 5 4 3 6 9 4 | | 5 4 3 8 9 4 |
| | 5 4 3 2 9 6 | | 5 4 3 4 9 6 | | 5 4 3 6 9 6 | | 5 4 3 8 9 6 |
| BM | 5 4 3 2 9 8 | BM | 5 4 3 4 9 8 | BM | 5 4 3 6 9 8 | BM | 5 4 3 8 9 8 |
| | 5 4 3 2 1 0 | | 5 4 3 4 1 0 | | 5 4 3 6 1 0 | | 5 4 3 8 1 0 |
| | 5 4 3 2 1 2 | | 5 4 3 4 1 2 | | 5 4 3 6 1 2 | | 5 4 3 8 1 2 |
| | 5 4 3 2 1 4 | | 5 4 3 4 1 4 | | 5 4 3 6 1 4 | | 5 4 3 8 1 4 |
| | 5 4 3 2 1 6 | | 5 4 3 4 1 6 | | 5 4 3 6 1 6 | | 5 4 3 8 1 6 |
| BM | 5 4 3 2 1 8 | BM | 5 4 3 4 1 8 | BM | 5 4 3 6 1 8 | BM | 5 4 3 8 1 8 |
| | 5 4 3 2 3 0 | | 5 4 3 4 3 0 | | 5 4 3 6 3 0 | | 5 4 3 8 3 0 |
| | 5 4 3 2 3 2 | | 5 4 3 4 3 2 | | 5 4 3 6 3 2 | | 5 4 3 8 3 2 |
| | 5 4 3 2 3 4 | | 5 4 3 4 3 4 | | 5 4 3 6 3 4 | | 5 4 3 8 3 4 |
| | 5 4 3 2 3 6 | | 5 4 3 4 3 6 | | 5 4 3 6 3 6 | | 5 4 3 8 3 6 |
| BM | 5 4 3 2 3 8 | BM | 5 4 3 4 3 8 | BM | 5 4 3 6 3 8 | BM | 5 4 3 8 3 8 |

Because our progression sequence was right to left, the first three progression sequences, a b and c , are the first three on the right, and the pattern on this page is duplicated on every page in the system. That is why the bdef parallel master key which we chose to illustrate parallel master keys controls the same positions on every page in the system.

Because our bdef parallel master key has four positions in common with the TMK, it will be capable of operating 256 combinations in a six pin system.

Remember from our basic masterkeying rules that if a key has one cut in common with the TMK, it will operate exactly 4 combinations in a two increment drop or 5 in a single increment drop. If it has two in common, 16 or 25 respectively. If it has three in common, 64 or 125 respectively. If it has four in common, 256 or 625 respectively. If it has five in common, 1024 or 3125 respectively. If it has six in common (on a seven pin) 4,096 or 15,625 respectively. Those rules do not change in dealing with parallel master keys. After all, those are merely incidental master keys, just like those which do appear on the charts.

There are 98 progression patterns for Incidental master keys on a seven pin lock cylinder, or 56 on a six pin. Seven plus the TMK are accounted for on standard six pin charts and eight on a seven pin chart.

They are: (Bold/underlined are on seven pin only )

(one pin masters - 4 combinations only )

a b c d e f **<u>g</u>**

( two pin masters - 16 combinations only )

ab ac ad ae af **<u>ag</u>**

bc bd be bf **<u>bg</u>**

cd ce cf **<u>cg</u>**

de df **<u>dg</u>**

ef **<u>eg</u>**

**<u>fg</u>**

( three pin masters - 64 combinations only )

abc abd abe abf **<u>abg</u>**

acd ace acf **<u>acg</u>**

ade adf **<u>adg</u>**

aef **<u>aeg</u>**

**<u>afg</u>**

bcd bce bcf **<u>bcg</u>**

bde bdf **<u>bdg</u>**

bef **beg**

**bfg**

cde cdf **cdg**

cef **ceg**

**cfg**

def **deg**

**dfg**

**efg**

( four pin masters - 256 combinations only )

abcd abce abcf **abcg**

acde acdf **acdg**

adef **adeg**

**aefg**

bcde bcdf **bcdg**

bdef **bdeg**

**befg**

cdef **cdeg**

**cdfg**

**cefg**

**defg**

( five pin masters - 1,024 combinations only )

abcde abcdf **abcdg**

acdef **acdeg**

**acdfg**

**adefg**

**bcdef** **bcdeg**

**bcdfg**

**cdefg**

( six pin masters - 4,096 combinations only )

abcdef **abcdeg**

**bcdefg**

The most commonly used of these are:

| | |
|---|---|
| def ( defg on 7 pin ) : | one combination per page, every page |
| adef ( adefg on 7 pin ): | one block per page, every page |
| abdef ( abdefg on 7 pin ): | one column per page, every page |
| ae ( ae on 7 pin ): | one block per page within each group in a section |
| abcef ( abcefg on 7 pin ): | one column per group, every group |

The following chart, though a bit bulky to use, shows exactly what each pattern would do, and can be carried with you or posted near the layout area for your convenience. To use it is actually quite simple. If the letter appears in the pattern, its condition is the upper of the pair. If it does not, it is the lower of the pair.

**YES**

| All 6 pin in 7 pin | All sections in 6 pin | All groups in section | All pages in group | All columns on page | All blocks in column | All combinations in block |
|---|---|---|---|---|---|---|
| **g** | **f** | **e** | **d** | **c** | **b** | **a** |
| One 6 pin per 7 pin | One section per 6 pin | One group per section | One page per group | One column per page | One block per column | One combination per block |

**NO**

For example, let's calculate our def from above. It does not have a, b or c, so those will be the lower of each pair, but it does have def, so those will be the upper. Therefore it will operate:

One combination per block
One Block per column
One Column per page
All Pages in group
All Groups in section
All sections in 6 pin

Now the bulky part which is simplifying what we learned. If it operates one combination per block, one block per column and one column per page, that means it really operates one combination per page, does it not? And similarly, if it operates all pages in group, all groups in section, and all sections in six pin, that is the same as saying every page in a six pin system. So what it really says is:
"One combination on every page of a six pin system".

Let's look at the adef. The only difference between it and the def is the addition of the a.

The a in the "Yes" means we use the upper instead of the lower, so instead of "one combination per block", it will be "all combinations in block." Therefore, instead of one combination per page, it will be one Block of combinations per page, and our def says it will be on every page in a six pin system

Easy, is it not? However, we usually will be working the other way, so let's look at that condition next.

We usually will be saying we need a combination that will meet a certain condition, such as all the combinations in a block, one block per group, every group in the system. Take a moment to try to calculate it before reading the solution.

Ready? All combinations in a block means "a" will be part of it, but we only want one block per group, so "b" and "c" will not. Position "d" says either all pages in group (Yes) or one page in group (No). Because we want only one block per group, that is a "No", so "d" will not be part of the pattern. However, we want it on every group in the system, so that means every group in a section and every section in a six pin system, so "e" and "f" will be "Yes", a part of the pattern. Thus the solution is "aef".

That was a lot more simple than you expected, was it not?

Remember that to use one of the parallel masters, all you need to do is look at your chart and decide if the level you need requires a master cut or an individual cut for each position to do what you want it to do. When you finish the positions, you will know the parallel master key you need for that special application within your system.

They are especially useful in overly complex situations such as Bill goes here, here, here and there, but Bob goes here, not there, here, not there...etc.

As an example of a parallel master key being used in a real world application, a hotel has 15 floors. Each pair of maids on a floor carries the master key for her floor only. However, there is a linen room on each floor that the laundry workers must access to pick up dirty linens and drop off clean in its place. The de parallel master key retains the necessary cut in position f to match each of the floor's individual keys, and has master cuts in positions d and e to pass all the floors, coupled with the individual key's cuts in positions a and b and c to give it its limitations. This is an elegant solution to avoid having to cross key the lock cylinders to add the selective master key.

And that is the true beauty of using parallel master keys...simple but elegant solutions to otherwise difficult situations. It is one amazingly useful tool in the Advanced Masterkeying Locksmith's virtual toolbox.

Here is the listing of what each does, although if you understand the chart above, you should never need this.

**a:** All combinations in block, One block per 6 pin or 7 pin

**b:** One combination per block, All blocks in column, One column per page, One page per 6 pin or 7 pin

**c:** One combination per block, One block per column, All columns on page, One page per 6 pin or 7 pin

**d:** One combination per page, All pages in group, One group per 6 pin or 7 pin

**e:** One combination per group, All groups in One section per 6 pin or 7 pin

**f:** One combination per section, All sections in One six pin per 7 pin

**g:** One combination per 6 pin, All 6 pins in 7 pin

**ab:** All combinations in column, One column per page, One page per 6 pin or 7 pin

**ac:** All combinations in block, One block per column, All columns on page, One page per 6 pin or 7 pin

**ad:** All combinations in block, One block per page, All pages in group, One group per 6 pin or 7 pin

**ae**: All combinations in One block per group, All groups in section, One section per 6 pin or 7 pin

**af:** All combinations in block, One block per section, All sections in 6 pin, One six pin per 7 pin

**ag:** All combinations in block, One block per 6 pin, All 6 pins in 7 pin

**bc:** One combination per block, All blocks in One page per 6 pin or 7 pin

**bd:** One combination per block, All blocks in column, One column per page, All pages in group, One group per 6 pin or 7 pin

**be:** One combination per block, All blocks in column, One column per group, All groups in section, One section per 6 pin or 7 pin

**bf:** One combination per block, All blocks in column, One column per section, All sections in 6 pin, One six pin per 7 pin

**bg:** One combination per block, All blocks in column, One column per 6 pin, All 6 pins in 7 pin

**cd:** One combination per column, All columns on page, All pages in group, One group per 6 pin or 7 pin

**ce:** One combination per column, All columns on page, One page per group, All groups in section, One section per 6 pin or 7 pin

**cf:** One combination per column, All columns on page, One page per section, All sections in 6 pin, One six pin per 7 pin

**cg:** One combination per column, All columns on page, One page per 6 pin, All 6 pins in 7 pin

**de:** One combination per page, All pages in One section per 6 pin or 7 pin

**df:** One combination per page, All pages in group, One group per section, All sections in One six pin per 7 pin

**dg:** One combination per page, All pages in group, One group per section, One section per 6 pin, All 6 pins in 7 pin

**ef:** One combination per group, All groups in section, All sections in 6 pin, One six pin per 7 pin

**eg:** One combination per group, All groups in section, One section per 6 pin, All 6 pins in 7 pin

**fg:** One combination per section, All sections in 6 pin or 7 pin

**abc**: All combinations on page, One page per 6 pin or 7 pin

**abd:** All combinations in column, One column per page, All pages in group,, One group per 6 pin or 7 pin

**abe:** All combinations in column, One column per group, All groups in section, One section per 6 pin or 7 pin

**abf:** All combinations in column, One column per section, All sections in 6 pin, One six pin per 7 pin

**abg:** All combinations in column, One column per 6 pin, All 6 pins in 7 pin

**acd:** All combinations in One block per column, All columns on page, All pages in group, One group per 6 pin or 7 pin

**ace**: All combinations in One block per column, All columns on page, One page per group, All groups in section, One section per 6 pin or 7 pin

**acf:** All combinations in One block per column, All columns on page, One page per section, All sections in 6 pin, One six pin per 7 pin

**acg:** All combinations in One block per column, All columns on page, One page per 6 pin, All 6 pins in 7 pin

**ade**: All combinations in One block per page, All pages in One section per 6 pin or 7 pin

**adf:** All combinations in One block per page, All pages in group, One group per section, All sections in 6 pin, One six pin per 7 pin

**adg**: All combinations in One block per page, All pages in group, One group per 6 pin, All 6 pins in 7 pin

**aef:** All combinations in One block per group, All groups in 6 pin, One six pin per 7 pin

**aeg**: All combinations in One block per group, All groups in section, One section per 6 pin, All 6 pins in 7 pin

**afg:** All combinations in One block per section, All sections in 6 pin or 7 pin

**bcd**: One combination per block, All blocks in group, One group per 6 pin or 7 pin

**bce:** One combination per block, All blocks in page, One page per group, All groups in One section per 6 pin or 7 pin

**bcf:** One combination per block, All blocks on One page per section, All sections in 6 pin, One six pin per 7 pin

**bcg:** One combination per block, All blocks on One page per 6 pin, All 6 pins in 7 pin

**bde:** One combination per block, All blocks in column, One column per page, All pages in One section per 6 pin or 7 pin

**bdf:** One combination per block, All blocks in column, One column per page, All pages in group, One group per section, All sections in 6 pin, One six pin per 7 pin

**bdg:** One combination per block, All blocks in column, One column per page, All pages in group, One group per 6 pin, All 6 pins in 7 pin

**bef:** One combination per block, All blocks in column, One column group, All groups in One six pin per 7 pin

**beg:** One combination per block, All blocks in One column per group, All groups in section, One section per 6 pin, All 6 pins in 7 pin

**bfg:** One combination per block, All blocks in One column per section, All sections in 6 pin or 7 pin

**cde:** One combination per column, All columns in section, One section per 6 or 7 pin

**cdf:** One combination per column, All columns in group, One group per section, All sections in 6 pin, One six pin per 7 pin

**cdg:** One combination per block, One block per column, All columns on page, All pages in group,, One group per section, One section per 6 pin, All 6 pins in 7 pin

**cef:** One combination per block, One block per column, All columns on page, One page per group, All groups in section, All sections in 6 pin, One six pin per 7 pin

**ceg:** One combination per block, One block per column, All columns on page, One page per group, All groups in section, One section per 6 pin, All 6 pins in 7 pin

**cfg:** One combination per block, One block per column, All columns on page, One page per group, One group per section, All sections in 6 pin, All 6 pins in 7 pin

**def:** One combination per block, One block per column, One column per page, All pages in group,, All groups in section, All sections in 6 pin, One six pin per 7 pin
deg: One combination per block, One block per column, One column per page, All pages in group,, All groups in section, One section per 6 pin, All 6 pins in 7 pin

**dfg:** One combination per block, One block per column, One column per page, All pages in group,, One group per section, All sections in 6 pin, All 6 pins in 7 pin

**efg:** One combination per block, One block per column, One column per page, One page per group, All groups in section, All sections in 6 pin, All 6 pins in 7 pin

**abcd:** All combinations in block, All blocks in column, All columns on page, All pages in group,, One group per section, One section per 6 pin, One six pin per 7 pin

**abce:** All combinations in block, All blocks in column, All columns on page, One page per group, All groups in section, One section per 6 pin, One six pin per 7 pin

**abcf:** All combinations in block, All blocks in column, All columns on page, One page per group, One group per section, All sections in 6 pin, One six pin per 7 pin

**abcg:** All combinations in block, All blocks in column, All columns on page, One page per group, One group per section, One section per 6 pin, All 6 pins in 7 pin

**abde:** All combinations in block, All blocks in column, One column per page, All pages in group,, All groups in section, One section per 6 pin, One six pin per 7 pin

**abdf:** All combinations in block, All blocks in column, One column per page, All pages in group,, One group per section, All sections in 6 pin, One six pin per 7 pin

**abdg**: All combinations in block, All blocks in column, One column per page, All pages in group,, One group per section, One section per 6 pin, All 6 pins in 7 pin

**abef:** All combinations in block, All blocks in column, One column per page, One page per group, All groups in section, All sections in 6 pin, One six pin per 7 pin

**abeg:** All combinations in block, All blocks in column, One column per page, One page per group, All groups in section, One section per 6 pin, All 6 pins in 7 pin

**abfg:** All combinations in block, All blocks in column, One column per page, One page per group, One group per section, All sections in 6 pin, All 6 pins in 7 pin

**acde:** All combinations in block, One block per column, All columns on page, All pages in group,, All groups in section, One section per 6 pin, One six pin per 7 pin

**acdf:** All combinations in block, One block per column, All columns on page, All pages in group,, One group per section, All sections in 6 pin, One six pin per 7 pin

**acdg:** All combinations in block, One block per column, All columns on page, All pages in group,, One group per section, One section per 6 pin, All 6 pins in 7 pin

**acef:** All combinations in block, One block per column, All columns on page, One page per group, All groups in section, All sections in 6 pin, One six pin per 7 pin

**aceg:** All combinations in block, One block per column, All columns on page, One page per group, All groups in section, One section per 6 pin, All 6 pins in 7 pin

**acfg:** All combinations in block, One block per column, All columns on page, One page per group, One group per section, All sections in 6 pin, All 6 pins in 7 pin

**adef:** All combinations in block, One block per column, One column per page, All pages in group,, All groups in section, All sections in 6 pin, One six pin per 7 pin

**adeg:** All combinations in block, One block per column, One column per page, All pages in group,, All groups in section, One section per 6 pin, All 6 pins in 7 pin

**adfg:** All combinations in block, One block per column, One column per page, All pages in group,, One group per section, All sections in 6 pin, All 6 pins in 7 pin

**aefg:** All combinations in block, One block per column, One column per page, One page per group, All groups in section, All sections in 6 pin, All 6 pins in 7 pin

**bcde:** One combination per block, All blocks in column, All columns on page, All pages in group,, All groups in section, One section per 6 pin, One six pin per 7 pin

**bcdf:** One combination per block, All blocks in column, All columns on page, All pages in group,, One group per section, All sections in 6 pin, One six pin per 7 pin

**bcdg:** One combination per block, All blocks in column, All columns on page, All pages in group,, One group per section, One section per 6 pin, All 6 pins in 7 pin

**bcef:** One combination per block, All blocks in column, All columns on page, One page per group, All groups in section, All sections in 6 pin, One six pin per 7 pin

**bceg:** One combination per block, All blocks in column, All columns on page, One page per group, All groups in section, One section per 6 pin, All 6 pins in 7 pin

**bcfg:** One combination per block, All blocks in column, All columns on page, One page per group, One group per section, All sections in 6 pin, All 6 pins in 7 pin

**bdef:** One combination per block, All blocks in column, One column per page, All pages in group,, All groups in section, All sections in 6 pin, One six pin per 7 pin

**bdeg:** One combination per block, All blocks in column, One column per page, All pages in group,, All groups in section, One section per 6 pin, All 6 pins in 7 pin

**bdfg:** One combination per block, All blocks in column, One column per page, All pages in group,, One group per section, All sections in 6 pin, All 6 pins in 7 pin

**befg:** One combination per block, All blocks in column, One column per page, One page per group, All groups in section, All sections in 6 pin, All 6 pins in 7 pin

**cdef:** One combination per block, One block per column, All columns on page, All pages in group,, All groups in section, All sections in 6 pin, One six pin per 7 pin

**cdeg:** One combination per block, One block per column, All columns on page, All pages in group,, All groups in section, One section per 6 pin, All 6 pins in 7 pin

**cdfg:** One combination per block, One block per column, All columns on page, All pages in group,, One group per section, All sections in 6 pin, All 6 pins in 7 pin

**defg:** One combination per block, One block per column, One column per page, All pages in group,, All groups in section, All sections in 6 pin, All 6 pins in 7 pin

**abcde:** All combinations in block, All blocks in column, All columns on page, All pages in group,, All groups in section, One section per 6 pin, One six pin per 7 pin

**abcdf:** All combinations in block, All blocks in column, All columns on page, All pages in group,, One group per section, All sections in 6 pin, One six pin per 7 pin

**abcdg:** All combinations in block, All blocks in column, All columns on page, All pages in group,, One group per section, One section per 6 pin, All 6 pins in 7 pin

**abcef:** All combinations in block, All blocks in column, All columns on page, One page per group, All groups in section, All sections in 6 pin, One six pin per 7 pin

**abceg:** All combinations in block, All blocks in column, All columns on page, One page per group, All groups in section, One section per 6 pin, All 6 pins in 7 pin

**abcfg:** All combinations in block, All blocks in column, All columns on page, One page per group, One group per section, All sections in 6 pin, All 6 pins in 7 pin

**abdef:** All combinations in block, All blocks in column, One column per page, All pages in group,, All groups in section, All sections in 6 pin, One six pin per 7 pin

**abdeg:** All combinations in block, All blocks in column, One column per page, All pages in group,, All groups in section, One section per 6 pin, All 6 pins in 7 pin

**abdfg:** All combinations in block, All blocks in column, One column per page, All pages in group,, One group per section, All sections in 6 pin, All 6 pins in 7 pin

**abefg:** All combinations in block, All blocks in column, One column per page, One page per group, All groups in section, All sections in 6 pin, All 6 pins in 7 pin

**acdef:** All combinations in block, One block per column, All columns on page, All pages in group,, All groups in section, All sections in 6 pin, One six pin per 7 pin

**acdeg:** All combinations in block, One block per column, All columns on page, All pages in group,, All groups in section, One section per 6 pin, All 6 pins in 7 pin

**acdfg:** All combinations in block, One block per column, All columns on page, All pages in group,, One group per section, All sections in 6 pin, All 6 pins in 7 pin

**acefg:** All combinations in block, One block per column, All columns on page, One page per group, All groups in section, All sections in 6 pin, All 6 pins in 7 pin

**adefg:** All combinations in block, One block per column, One column per page, All pages in group,, All groups in section, All sections in 6 pin, All 6 pins in 7 pin

**bcdef:** One combination per block, All blocks in column, All columns on page, All pages in group,, All groups in section, All sections in 6 pin, One six pin per 7 pin

**bcdeg:** One combination per block, All blocks in column, All columns on page, All pages in group,, All groups in section, One section per 6 pin, All 6 pins in 7 pin

**bcdfg**: One combination per block, All blocks in column, All columns on page, All pages in group,, One group per section, All sections in 6 pin, All 6 pins in 7 pin

**bcefg:** One combination per block, All blocks in column, All columns on page, One page per group, All groups in section, All sections in 6 pin, All 6 pins in 7 pin

**cdefg:** One combination per block, One block per column, All columns on page, All pages in group,, All groups in section, All sections in 6 pin, All 6 pins in 7 pin

**abcdef:** All combinations in block, All blocks in column, All columns on page, All pages in group,, All groups in section, All sections in 6 pin, One six pin per 7 pin

**abcdeg:** All combinations in block, All blocks in column, All columns on page, All pages in group,, All groups in section, One section per 6 pin, All 6 pins in 7 pin

**abcdfg:** All combinations in block, All blocks in column, All columns on page, All pages in group,, One group per section, All sections in 6 pin, All 6 pins in 7 pin

**acdefg:** All combinations in One block per each column on All pages in 7 pin

**bcdefg:** One combination per block, All blocks in column, All columns on page, All pages in group,, All groups in section, All sections in 6 pin, All 6 pins in 7 pin

**abcdefg:** All combinations in block, All blocks in column, All columns on page, All pages in group, All groups in section, All sections in 6 pin, All 6 pins in 7 pin

This page left intentionally blank.

# Chapter Two:

# Master Key Application

**The best laid out Standard Progression Format chart is useless if it cannot be applied to the customer's system. That cannot happen unless you understand the customer's needs within the system.**

The application phase consists of:

- Key system meeting
- Examining the system
- (Floor plans or walk-through)
- Rough layout (Grid chart and SKC)

**Each of the steps in this phase is a critical one. No one can build a great (meaning really good-not referring to a level of keying) master key system if the information they have going into it is flawed. Take your time with each step.**

Often you will be dealing with people who are on tight schedules and need to complete this phase as quickly as possible. Remember that it will take less of their time (and yours) to do it right at this phase of the project.

There is an old saying:

**Always rush slowly and carefully**

**(There is no job that cannot be done faster by doing it right the first time.)**

# Evaluating Customer Needs

### (Prior to the Key Meeting)

## Departmental Structure

- How many departments in the building?
  - This will give you an indication as to how many "sub-master key" groups you will need to break the system into for the building.
- Do they share some spaces?
  - These spaces will probably require cross-keying or special keying of some kind to fulfill the needs of both departments. This might also indicate a need for a function oriented areas where a selective master key might be useful.
- Are there spaces exclusively for higher level employees?
  - These spaces will almost certainly require special keying such as keying directly under a particular master key.
- Are there spaces primarily for lower level employees?
  - Once again this may indicate a need for special keying such as Selective Master Keying.
- Do they sign out temporary keys?
  - This may indicate a need to break the system into smaller Master Key groups so that lost keys will be a minimal risk and expense.

## Special Assignments

- Are there areas not opened by master keys?
  - This indicates a need to allow for sufficient extra room for expansion to pull SKD combinations with.
- Are there areas opened by all keys?
  - This indicates a definite need for cross keying, most likely by leaving selected chambers empty in order to defeat Dayton Method attacks.
- Are there areas requiring limited extra access?
    - These may indicate a need to use parallel Master Keys or the special Master Keys from an SPF chart.

- **High Security Areas**
  - Are there areas such as cash counting rooms that need higher security?
    - These may indicate a need for SKD or Keyed Directly Under a particular higher level Master Key.
  - Are there areas where robbery or kidnapping is a threat?
    - These may indicate a need for SKD or Keyed Directly Under a particular higher level Master Key.
  - Are there secluded areas where attacks could occur?
    - These may also indicate a need for SKD or Keyed Directly Under a particular higher level Master Key.

Taking some time to develop answers on your own prior to the key meeting can make you seem more like an expert during the meeting and help you to maintain your level of control over the meeting.

If you are not able to establish that this part of the meeting is "yours", it is likely to deteriorate into silly discussions such as:

"Yeah, but what about Bob? He sometimes comes early and would want to use that door because it is closer to the parking lot. But I am not sure I feel comfortable with him having that key."

Obviously, this type of question should have been answered earlier, or can wait until key issue procedures are developed when the system is being installed. If you are in charge, you can keep from becoming bogged down in such details that do not really affect the system application.

# Running an Effective Key System Meeting

(Immediate Considerations)

- *Are you talking to the correct individual?*
  - Do they have the authority to actually make/approve decisions?
  - Too little authority makes meetings a waste of time
  - Are they familiar with the people and the system needs?
  - Often too high of authority can be as bad as too little.
  - Do they understand the terminology and restrictions?
    - This is where supplying them with general information on terminology and lockset functions in advance can save a lot of time and make meetings more productive.
  - Have they discussed the system layout with those affected by it?
    - Encouraging an increase the level of communication at the early stages can save a lot of problems later on.
    - The time they spend now with their staff will be less time-consuming than dealing with all the complaints once the system has been developed.

# Running an effective key system meeting:
## (The actual conduct of the meeting)

- If at all possible, provide each attendee with a list of doors/rooms and hardware functions prior to the meeting. Ask that they review it with their key personnel to determine who actually should have access to each area.

    - This will improve their level of knowledge about the "real" systems in place currently. Many of the department heads, etc., have little or no contact with the lower echelon functions involved in their operation, and may not be fully aware of the needs of system users. Providing them with the opportunity to review it with their people in advance of the meeting will save a multitude of "false starts" at and following the meeting.

- Remember that although we will be using the people's need for access as a tool to help us lay out the system, we are:

**"Keying the Facility", not “Keying the People”.**

Although seeing who needs access to which areas is an effective tool in determining how to effectively group access, there is a tendency to look at it as though each person "needs" to carry only one key, which is neither practical nor possible in most situations.

- Understand that there are finite limitations on the levels of keying which are possible.

- Decisions as to adding a level or increasing the number of combinations in a level may affect your ability to deliver the system as promised.

- Look for an overview of the system before getting into the "door by door" itemization. The understanding of logical divisions within the facility is vital to effective key system layout.

- Look for departmental areas, areas of shared responsibilities or resources, and areas of extremely limited usages.

- Grouping areas by function or availability of access is usually logical, but don't make assumptions without following up on them with your customers.

- If possible, work on the system "From the Top Down" first, and then review it for feasibility "From the Bottom Up".

Laying the initial system out from a "Top Down" perspective allows for logical groupings in most cases. However, when you check the system for feasibility it is vital to work "From the Bottom Up", because it is the numbers of combinations at each of the lower levels that are most likely to limit your system.

Understand the impact that the Manufacturer's key system and tolerances has on system layout.

Most manufacturers of commercial locksets offer a six pin "Base Four" system, where each area breaks up into potential logical groups of "fours".

A manufacturer whose system instead uses a "Base Five" system, increases the available combinations and levels twenty-five per cent, while a manufacturer whose tolerances demand a three step increment or who only offers a five pin lock cylinder reduces the number of useful
and available combinations by as much as 75 per cent!

A manufacturer offering a seven pin cylinder (such as is common on many Interchangeable Core products) increases the number of combinations or levels by a multiplier similar to the "System Base" figure. If it is a Base Four system, it is increased by 400 %, for example, or 500% for a Base 5 system.

# Special Applications and Needs

OK, are you ready for one of Master Keying's best kept secrets? Even though we have been talking about incidental master keys in the traditional sense that we think of as master keys, an incidental master key of any level can be used as a change key for your customer's system when it is appropriate to do so to satisfy the needs of the system.

For example, what if we had a situation where two keys from the same general master key group had to operate a cross keyed cylinder. Instead of stacking pins to accomplish it, what if we used, for example a line master key as one change key and a block master key as another change key, crossing off the rest of the combinations they would have controlled.

Once again, this provides for the needs of the system without a reduction in the physical security of the cylinder.

There are many applications where an incidental master key (parallel or not) can be used effectively as a change key. This is fine as long as you remember to cross off the other combinations the master would have operated that will not now be usable.

It is actually amazing how many locksmiths have not recognized that simple fact.

The initial layout of a system is generally fairly straight forward and simple. We start at the lowest level of master key in the system and work "up", determining the appropriate levels as we go.

The reason we work "up" the charts and not down is a simple one to see.

Let's say we have a master key that operates 30 garden style residences in a single increment system (where everything breaks up into groups of fives) where there are four private sleeping rooms in each apartment. If there is no outer door bitting and the inner doors pass the outer door lock, it only yields a need for 120 cylinder bittings.

Ignore the need for future expansion for now. A single page of our SPF for the single increment drop (Base Five) gives us 125 change key bittings, which should be more than enough, so a Page Master is sufficient, right?

Wrong!

Because we will not want to stack a series of number one master pins in one chamber, it is logical to assume that we will leave one position (progression sequence a) empty in the outer apartment door lock, and choose the four combinations from one block for each apartment. That means we are reserving 30 blocks. There are only 25 blocks on a page. Therefore a Page Master will NOT be sufficient!

We were misled by working "down" the chart, i.e., looking at how many change keys fell under a higher level master instead of what groupings must fall under it. We find the groupings by working "up" the chart...calculating the needs of the lower levels first and then comparing them for the needs of the upper level master.

Had we done this on the example here, we would have said that each room comes out of a Block of change key bittings, and that the four rooms basically wipe out the block, especially if we are leaving an empty chamber in the outer apartment door cylinder. Therefore we need a Block reserved for each of the thirty apartments, or thirty blocks. There are only 25 on a page, so the master above them cannot be a Page Master but requires a Group Master instead.

Let's look at another misleading example:

We have a two increment drop (Base Four) system on a five pin lock cylinder with five master keys of 17 change keys under each under a grand master key and a great grand master key. The second grand master key has only two master keys of 5 change key bittings each.

Total change key cuts: 95

So that means a Group Master capable of handling 256 change keys should be able to handle the whole thing, right?

If you did not hear this "Wrong" coming, you have not been paying attention!

Working from the bottom up, we see that 17 change keys is too much for a Vertical or Horizontal Column Master to handle, so each requires a Page Master.

And there are FIVE of them in a system with a multiplier of FOUR, so that means the next level up, the Group Master, cannot handle the five masters, so it instead must be the Section Master, which uses five cuts from the TMK in sequence

positions a, b, c, d, and e, which is all the room there using five pin cylinders, leaving no room to pull the grand master with two master keys of five each from!!! The system cannot be accomplished using Total Position Progression (TPP) on a five pin key. Either the system, the method or the lock cylinders have to change, or this job cannot be done!
Notice that we did not say the job cannot be done at all. If we change the system requirements to not have as many master key groups, it can be done. If we change the method to use Rotating Constant Method (RCM), it can be done. If we change the cylinders to six pin (or even better yet, to seen pin), it can be done.

This is why it is important to hold the keying meeting with the system owners early in the process, and do the layout almost immediately, while there is still time to make corrections such as this.

But what about other special needs?

# Special Situations

**No Grand Master Key (2 level)**:
Under the SKCS labeling, the master key typically is called AA, and the change keys are 1AA, 2AA, 3AA, etc. The numbers are PREFIXED TO ( added to the front of ) the master key symbol. This is a Level 2 System. Standard bittings from the charts will usually operate these.

**KEYED DIRECTLY TO:**
A cylinder or cylinders are operated by no keys lower than the specified master key level. Standard coding takes the symbol of that Master Key Level, such as A for Keyed Directly To the Grand Master Key. We usually pin these by calculating the TMK cuts and this level of master key's cuts.

**KEYED DIRECTLY UNDER:**
A cylinder or cylinders are operated by individual keys and the specified Master Key level, but by no intermediate levels. Using the SKCS labeling, these take the symbol of the specified Master Key Level plus a number Suffixed to it, (A1,A2, B1). We typically pin these

using an unused portion of the chart the master directly relates to.

**SINGLE KEYED:**
No Master Key of any level operates this (group of) cylinder(s). Takes the symbol SKD suffixed by a number, such as SKD1,SKD2. They may be a totally different keyway, or may be an unused bitting from the charts. It is usually pinned directly to bottom pins matching the cuts on the key involved, with no master pins.

**NMK: No Master Key:**
This is another less common designation for SKD, that is used when the SKD bitting will also be an active bitting elsewhere in the system, where it WILL fall under other master key(s). It is usually pinned directly to bottom pins matching the cuts on the key involved, with no master pins, in the lock cylinders where it is NMK and pinned in the traditional manner of master keyed cylinders elsewhere in the system.

**MAISON KEYED:** Keyed so that all the keys from an area, such as an apartment building, will operate it. This is actually not considered a proper term in Master Keying, but it is in such common usage, that it deserves to be included in this list.

You may see it also listed as MAISSON KEYED. Both terms are actually correct, being taken from the French for 'house'. Classic French (now long obsolete) used the Maisson as an informal alternative to 'manoir', while the 'maison' indicated a more traditional house. Today's French, however, makes no such distinction and the term 'Maisson' has dropped from usage entirely.

# Cross Keying

**In some circumstances you may have two or more change keys that have to pass a given lock. This is called Cross Keying. A cross keyed lock can have its own key or not.**

Cross keying may be considered Controlled Cross-Keying or Uncontrolled Cross Keying.

These terms may be a bit deceptive, because they have nothing to do with whether you intended it or not. "Cross-Keying" that you did not intend to happen is called **INTERCHANGE** and is a bad thing.

## CONTROLLED CROSS KEYING:

**A condition in which two or more different keys of the same level of keying and under the same higher level key(s) operate the cylinder by design; e.g., XAA1 operated by AA2 (but not XAA1 operated by AB1)**

## UNCONTROLLED CROSS KEYING:

**A condition in which two or more different keys under different higher level keys operate one cylinder by design; e.g., XAA1 operated by AB, AB1**

The terms Controlled and Uncontrolled refer to the fact that in pinning the cylinder, pinning a controlled cross-keyed cylinder tends to be more straightforward than an uncontrolled, and it is easier to be certain that none of the keys that are NOT intended to operate the cylinder will in fact operate it.

# Cross Keying may also be with or without its own key:

### Cross Key with its own key

- AA1 works the main door but not the inside offices.
- AA2, AA3 and AA4 work their own offices and the main door

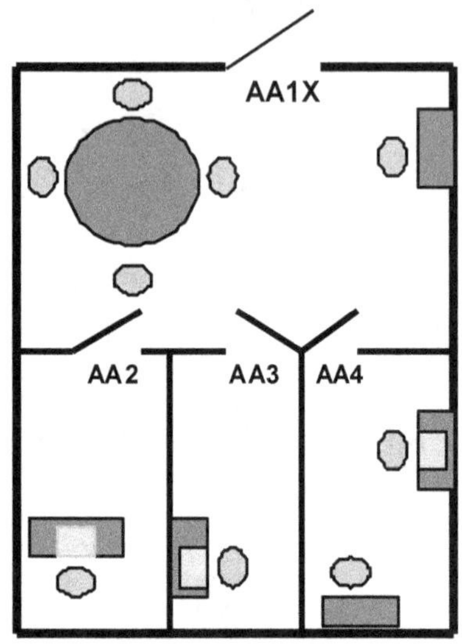


**CROSS KEYED WITH ITS OWN KEY**: Prefixed with X in Front of Standard Symbol for its key ( XAA1, XBC5, etc. )

**CROSS-KEYING WITHOUT ITS OWN KEY:** Pair of X symbols separated by a Number (X1X, X2X,etc.) Some beginner master keying locksmiths make the mistake of calling this X I X, as though it were the letter i instead of a number. Don't be one of them.

# Cross Key without its own key

**Cross Keying without its own key:**

- AA1, AA2 and AA3 all work their own rooms but also all work X1X also.
- There is no individual key for doors X1X.

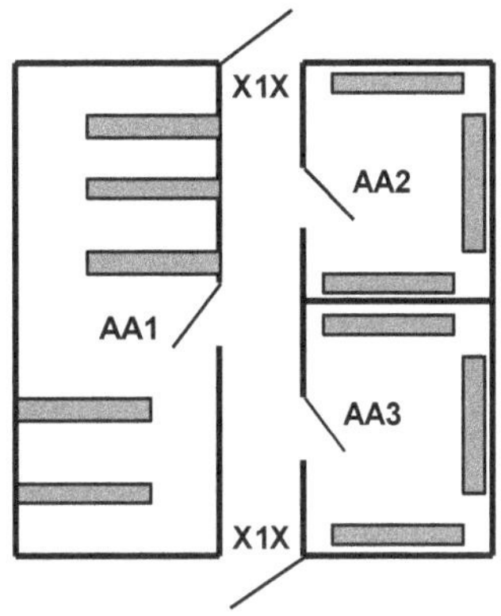


**Remember:**

Says ***STOP!*** There is something unique in my keying!

**Only by looking at the notes can you see what the unique keying is for an "X" marked lock, under the SKCS.**

**For example, the notes may say:**

**A1X operated by AA2, AA3, AA4**

**X1X operated by AA1, AA2, AA3**

**AB1X operated by AB1 through AB64**

An abbreviation used by some manufacturers is:

"O.b" to stand for "Operated by"

For example:

***AA1X O.b AA2, AA3, AA4***

***X1X O.b AA1, AA2, AA3***

AB1X O.b AB1 through AB64

Almost every special situation can be easily handled with just a little thought and planning, but the cross keying is the most complex. We do NOT want to "stack master pins" because this frequently causes many malfunctions and problems.

In the previous chapter, you learned about the use of parallel master keys. In this chapter you have seen that master keys of any level can be used as though they were change keys by using them for situations such as cross keying. Remember, though, to cross off the remaining keys under their authority if you do so.

When we look at the Rotating Constant Method (RCM) you will find even more possibilities of ways to handle special circumstances.

The key to every special situation lies in your careful planning and layout, remembering that the SPF charts are a useful guideline on how to apply the system, not an immutable and fixed law to limit you. Use your logic and understanding. Know that your knowledge of the options and possibilities is the real answer as to how to select the best bittings and options for the system.

While not a tool you will use every day or in every system, this knowledge is a valuable tool in the Advanced Master Keying Locksmith's toolkit.

# CHAPTER THREE:
# THE LIMITED PROGRESSION TECHNIQUE

### (previously known as the Partial Progression Technique)

***NOTE: This chapter is copied directly from the Master Keying Text Book, so if you are fully familiar with it, you can skip this chapter and go on to Selective Master Keying.**

## (A subset of RCM that can be developed using TPP charts)

In the previous section, we saw that by pinning our cylinders to the Change Key combination and the bitting of the TMK, each 6 pin cylinder had 62 incidental keys that would also operate it IF THEY EXISTED. If it was a five pin cylinder, each would have 30 incidental master keys that would also operate it IF THEY EXISTED. If it had been a 7 pin cylinder, each would have been 126 incidental master keys per cylinder.

BUT, by assigning them to jobs as master keys of various levels, we were able to do away with the problems they pose. Right?

WRONG!!! This would only be true if the customer's system were so complex as to require each and every one of them. We have preserved security by NOT INCREASING the number of incidental keys, (as would happen if we stacked pins), but we have only DECREASED their number by any master keys that we ACTUALLY NEEDED for the system!!!

If, for example, the customer's system only required a Grand Master Key and 2 Master Keys under it, each cylinder would need to be operated by only its change key, the highest master key in the system, and ONE LOWER LEVEL MASTER KEY! So instead of having 62 phantom keys, we have 61. So what? That's not much of an increase in security!!!

So what is the solution? Are we locked into the loss of security? In most cases the answer is "NO"!!!

**The answer lies in selecting APPROPRIATE pinning levels for the specific job at hand, including potential needs for anticipated or reasonably possible expansion.**

If, for example, we needed only 4 change keys controlled by one master key, it would be folly indeed to use as our master the TMK!!! Instead, in a system where the KBA has 4 or 5 possibilities in the a position or c position, a BLOCK MASTER or a LINE MASTER (respectively) might be chosen to be the highest master key actually required by the system.

**And if we're not going to use the higher master, why pin to it?**

**Instead, we pin only to the level of master key actually required by our system.** For example, using the Sample chart we worked on earlier, we see that the Block Master for the first block on page one was 5 4 3 2 7 8, so our pinning would be:

| | |
|---|---|
| BMK | 5 4 3 2 7 8 |
| CK | 5 4 3 2 7 0 |
| Smallest=Bottom Pin | 5 4 3 2 7 0 |
| Difference=Master Pin | * * * * * 8 |

Can you determine how many incidental master keys exist between the BMK and the Change Key in this cylinder?

If you determined the BMK to be the only master key of any level in that cylinder, you are correct.

That will not be the case every time this technique is applied, of course. The higher the key you choose to use as your Actual Top Master Key, the more Incidental Master Keys will remain.

**For example, if we chose the VMK for our actual Top Master, that cylinder's pinning would be:**

| | |
|---|---|
| VMK | 5 4 3 2 5 8 |
| CK | 5 4 3 2 7 0 |
| Smallest=Bottom Pin | 5 4 3 2 5 0 |
| Difference=Master Pin | * * * * 2 8 |

This would permit Incidental Master Keys (Don't ask why the abbreviation IMK has never caught on.) of 543250 and 543278 (the BMK) to operate in addition to the two keys we planned.

**For example, if we chose the PMK for our actual Top Master, that cylinder's pinning would be:**

| | |
|---|---|
| PMK | 5 4 3 0 5 8 |
| CK | 5 4 3 2 7 0 |
| Smallest=Bottom Pin | 5 4 3 0 5 0 |
| Difference=Master Pin | * * * 2 2 8 |

This would permit Incidental Master Keys of 543250, 543050, 543258 (the VMK), 543078 (the HMK), 543070 (the Line Master) and 543278 (the BMK) to operate in addition to the two keys we planned.

**Notice that any designated Incidental Master Keys below the actual chosen Top Master will remain in operation, but all Masters above it are removed from operational functionality. Now you see why this book occasionally referred to the TMK as the Theoretical Master Key. When using this technique, the TMK may actually be taken out of the system and the system broken into many smaller systems instead.**

**REMEMBER: Pinning it to a higher level of master key automatically allows the lower levels of master key under it to also work their areas.**

**As you have just seen, we can pin the lock cylinders to any level of master key we need to. The fundamental rule is that we pin it to the HIGHEST LEVEL OF MASTER KEY ACTUALLY REQUIRED BY THE SYSTEM.**

For now, let's look at its effect on security!!!

**If we pin to a Block Master (4 or 5 Change Keys)** as the highest master key in the system, how many incidental master keys exist for the lock cylinders involved? NONE!!! That's right, not a single unintended key will open any of those four locks. From 62 down to ZERO!!!

**If we pin to a Vertical Column Master or a Horizontal Master (16 or 25 Change Keys) as the highest master key** for the system, there are 2 unintended keys, and one of them is the Block Master. That's a BIG improvement.

**If we pin the Page Master (64 or 125 Change Keys) as our highest master key in the system**, each lock cylinder will have 6 unintended keys to operate it. From 62 to 6 is still excellent! Furthermore, one of those 6 is the Block Master, one is the Horizontal Master, one is the Horizontal Master and one is the Vertical Master.

**Pinning to a Group Master (256 or 625 Change Keys) as the highest master key** will allow 14 unintended keys to operate any given cylinder. From 62 to 14 is still a good increase in security, but one of the 14 is a Block Master, one is a Line Master, one is a Vertical Master, and one is a Horizontal Master, and one is the Page Master. This could conceivably decrease the number of unintended keys to as few as 9.

**Pinning to the Section Master (1,024 or 3,125 Change Keys)** doesn't give us as much an increase in security as the lower levels. It reduces it to 30, which is still about a 50% increase in security! Once again, one is the Block Master, one the Line Master, one the Vertical Master, one the Horizontal Master, one the Page Master, and one the Group Master (Base Master).

**Whenever we pin every cylinder directly to a Change Key and any level of master key other than the theoretical master, we have created a system using the technique of Limited (formerly Partial) Progression!!!**

**It should be noted that this is actually a subset of the Rotating Constant Method of Master Keying, which should be understood before employing the technique on existing systems.**

**It should not be used to "key into existing systems" unless the entire system including all expired or changed bittings is known. (The following chapters will explain why.)**

# Chapter Four: Selective Masterkeying

Incidental Master Keys have been the standard since their use was strongly introduced and promoted by Masterkeying greats **Wiegend Jensen** and **Bert Michaels** in 1976, and the concept was a major part of MK101 developed in 1982 by the ALOA Sponsored National Task Group for Certified Training Programs, Master Keying Study Group, which later became the LIST (Lock Industry Standards and Training) Council. Although the concept dated back to before 1960 it had been largely ignored until Bert Michaels and Wiegend Jensen spread it throughout the industry and made it simple enough for the average locksmith of the time to easily grasp, and the LIST Council made it a basic concept for locksmiths.

Since that time, incidental master keys have been the standard, and "**stacking pins**" to achieve levels of master keys has been strongly discouraged. However, the truth is that in some systems, a greater level of security and flexibility can be gained by "stacking pins" rather than relying on incidental masterkeys at higher levels.

This technique is called **Selective Masterkeying**, and (with the exception of parallel master keys as seen in the previous chapter) typically relies on the use of "**Unassociated Master Keys**"- keys which are not Incidental to the system.

The concept of Selective Masterkeying is that it makes sense to use Incidental Master Keys for lower level master keys that can easily be re-keyed if they are compromised, but that in a large system, the costs associated with rekeying if a high level master key is compromised might be overwhelming, perhaps so much that the rekeying does not occur as it needs to.

Take, for example an actual mid-sized University in a largely suburban setting, with 33 traditional residence halls (not counting Sororities or Fraternities), ranging in size from 42 students to 250. Each residence hall is divided into control areas based on wings of the building, so the number of students affected by a compromised lower level master key ranges from 16 to 50. While changing lock combinations and issuing new keys to 50 students is a difficult task, it is not an overwhelming one, and doing the same for 16 students is an easy task.

But now think of the time, expense and difficulties of issuing over 3000 students new keys and rekeying all of their lock cylinders because a top level master key was compromised. This University was chosen for an example because it had to do that, and did. The cost was close to a **million dollars** with a large portion of that in materials costs, and it took three weeks to fully implement. But it was necessary because the top master key was the basis of the entire incidental-based masterkey system throughout the residence halls.

Seven years later the same level of master key was lost. The cost, including mostly labor, was **under $30,000.** The difference was that when the system was replaced it instituted a Selective Master key system. **Because there was not a direct relationship between the individual keys and the high level master key, the high level master key could be replaced without the need to issue new individual keys or intermediate level master keys.** The locks were all rekeyed, of course, but it was accomplished in just one week.

So let's look at this miracle technique. The truth is that it is extremely simple. You design the system as though you were designing an incidental master key system, because you are - for the lower levels of master keys. But remember that the higher levels are only incidental because you have pinned them that way - using Total Position Progression and pinning all six (or seven) chambers accordingly. What if instead you had used the Limited Progression technique and pinned only three of the chambers to include master pins? No Master Key higher than a Page Master Key would operate any of the cylinders. Correct?

That would satisfy most system needs, but if you had areas with more locks than that you could instead still pin only four chambers to accept master pins, limiting the highest working Master Key to that of a Group Master Key, and still limiting the number of locks that would need rekeyed if that level of master key were compromised.

But what about custodial, maintenance, HVAC and Campus Police? They all need into most of the rooms or all of the rooms. But do they really need into all of the rooms all of the time?

Let's face it, custodial is not generally providing a true emergency service, even though a spill needing mopped up might pose a threat to life safety. Generally they are assigned to a building or small group of buildings for their entire shift. Why not set up a secure location where they can be issued the Incidental Master Keys they require for the day, and let them carry a few if need be?

Similarly, unless you live in the Desert or at Frigid Falls, you probably do not have a truly life-threatening emergency for HVAC. Could they also be issued the actual keys they require for the day or for an individual job or group of jobs? They may not like it, but the truth is it makes sense.

That being said, we still have maintenance and Campus police, who it can be reasonably well argued do have a need for the protection of life to have quick and ready access to all areas. So for them we add an individual master pin, preferably a three or higher in length. This creates a key that they consider the top master key. However, if it is compromised, it takes only changing that one pin to remove that top master key and create a different one in the chambers we did not initially masterkey.

So what if custodial or HVAC still needs a master key for access? We have two choices. We can give them the Police/Maintenance master key, thus reducing its effectiveness, or we can simply add a separate master pin in one of the un-mastered chambers for them. That way if they lose their top master key, it can be rekeyed to a new one without affecting emergency services. In effect, we have multiple "top master keys".

Using the chart shown below, let's look at how this might be applied. We will assume that we are looking at the page representing an average size building with four wings, which we will call A1, A2, B1 and B2.

---

| f | e | d | c | b | a | |
|---|---|---|---|---|---|---|
| 3 | 2 | 1 | 0 | 5 | 8 | THEORETICAL MASTER |
| 5 | 4 | 3 | 2 | 7 | 0 | BASE FIELD |
| 7 | 6 | 5 | 4 | 9 | 2 | (KBA) |
| 9 | 8 | 7 | 6 | 1 | 4 | KEY BITTING ARRAY |
| 1 | 0 | 9 | 8 | 3 | 6 | |

f e d c b a ⇐ SEQUENCE OF PROGRESSION

PAGE MASTER ______ PAGE # 1

GROUP MASTER ______ OF 64 PAGES

SECTION MASTER ______

| | # | VERTICAL MASTER | # | VERTICAL MASTER | # | VERTICAL MASTER | # | VERTICAL MASTER | LINE MASTERS ⇩ |
|---|---|---|---|---|---|---|---|---|---|
| | 1 | 5 4 3 2 7 0 | 17 | 5 4 3 4 7 0 | 33 | 5 4 3 6 7 0 | 49 | 5 4 3 8 7 0 | |
| | 2 | 5 4 3 2 7 2 | 18 | 5 4 3 4 7 2 | 34 | 5 4 3 6 7 2 | 50 | 5 4 3 8 7 2 | |
| HORIZONTAL MASTER | 3 | 5 4 3 2 7 4 | 19 | 5 4 3 4 7 4 | 35 | 5 4 3 6 7 4 | 51 | 5 4 3 8 7 4 | |
| | 4 | 5 4 3 2 7 6 | 20 | 5 4 3 4 7 6 | 36 | 5 4 3 6 7 6 | 52 | 5 4 3 8 7 6 | |
| | | BLOCK MASTER | | BLOCK MASTER | | BLOCK MASTER | | BLOCK MASTER | |
| | 5 | 5 4 3 2 9 0 | 21 | 5 4 3 4 9 0 | 37 | 5 4 3 6 9 0 | 53 | 5 4 3 8 9 0 | |
| | 6 | 5 4 3 2 9 2 | 22 | 5 4 3 4 9 2 | 38 | 5 4 3 6 9 2 | 54 | 5 4 3 8 9 2 | |
| HORIZONTAL MASTER | 7 | 5 4 3 2 9 4 | 23 | 5 4 3 4 9 4 | 39 | 5 4 3 6 9 4 | 55 | 5 4 3 8 9 4 | |
| | 8 | 5 4 3 2 9 6 | 24 | 5 4 3 4 9 6 | 40 | 5 4 3 6 9 6 | 56 | 5 4 3 8 9 6 | |
| | | BLOCK MASTER | | BLOCK MASTER | | BLOCK MASTER | | BLOCK MASTER | |
| | 9 | 5 4 3 2 1 0 | 25 | 5 4 3 4 1 0 | 41 | 5 4 3 6 1 0 | 57 | 5 4 3 8 1 0 | |
| | 10 | 5 4 3 2 1 2 | 26 | 5 4 3 4 1 2 | 42 | 5 4 3 6 1 2 | 58 | 5 4 3 8 1 2 | |
| HORIZONTAL MASTER | 11 | 5 4 3 2 1 4 | 27 | 5 4 3 4 1 4 | 43 | 5 4 3 6 1 4 | 59 | 5 4 3 8 1 4 | |
| | 12 | 5 4 3 2 1 6 | 28 | 5 4 3 4 1 6 | 44 | 5 4 3 6 1 6 | 60 | 5 4 3 8 1 6 | |
| | | BLOCK MASTER | | BLOCK MASTER | | BLOCK MASTER | | BLOCK MASTER | |
| | 13 | 5 4 3 2 3 0 | 29 | 5 4 3 4 3 0 | 45 | 5 4 3 6 3 0 | 61 | 5 4 3 8 3 0 | |
| | 14 | 5 4 3 2 3 2 | 30 | 5 4 3 4 3 2 | 46 | 5 4 3 6 3 2 | 62 | 5 4 3 8 3 2 | |
| HORIZONTAL MASTER | 15 | 5 4 3 2 3 4 | 31 | 5 4 3 4 3 4 | 47 | 5 4 3 6 3 4 | 63 | 5 4 3 8 3 4 | |
| | 16 | 5 4 3 2 3 6 | 32 | 5 4 3 4 3 6 | 48 | 5 4 3 6 3 6 | 64 | 5 4 3 8 3 6 | |
| | | BLOCK MASTER | | BLOCK MASTER | | BLOCK MASTER | | BLOCK MASTER | |

---

Each Wing, having less than 16 locks on it, gets a Vertical Master Key, directly from the chart, to be held in the staff office for the Residence Hall Assistant (RA or Gryphon) to use to handle lockouts and minor emergencies. The Residence Life Coordinator (or Hall Director) gets the Page Master Key directly from the chart.

But instead of having the remaining three chambers pinned to a TMK, we are creating two separate "Top Master Keys"-one for Campus Police and Maintenance, and one for HVAC and Custodial.

Our Campus Police/Maintenance Top Master Key will be 1 4 3 0 5 8 and the Custodial/HVAC Top Master Key will be 5 0 3 0 5 8.

The Pinning will thus be:

| | | |
|---|---|---|
| CP/Maint | 1 4 3 0 5 8 | |
| Cust/HVAC | 5 0 3 0 5 8 | |
| Hall Dir | 5 4 3 0 5 8 | |
| Ind Key | 5 4 3 2 1 0 | ( first combination in column ) |
| BP | 1 0 3 0 1 0 | |
| MP | 4 4 - 2 4 8 | |

Note that in this example, we assumed zero to be a shallow cut. Notice also that because in this example we did not need to use four chambers for the Hall Director's Master Key, we ended up with more security with no master pin in the third chamber from the face. Remember that this will not always be the case.

If the Campus Police/Maintenance key is compromised, we have many other combinations available to replace it, such as 343058, 743058. 547058, etc.

Of course, if your system is 7 pin you can gain a great deal more security with this method without the limitations it imposes on a six pin system.

Another option is to not use the two increment drop possibilities for the selective master keys, but to use a three increment drop instead for them. This will limit the number of possible selective master keys available but will increase the physical security and reduce the likelihood of master pin "slippage" (from one position within the lock to another) or "jamming" (turning sideways in the chamber and preventing proper key operation.)

This is just one example of how the selective master key technique can be applied and used as an effective tool in the Advanced Master Keying Lockmith's toolkit.

# CHAPTER FIVE

# THE ROTATING CONSTANT METHOD (RCM)

As we saw earlier, we can develop a system using Right to Left Progression or by varying the progression sequence. Varying it changed the order in which each combination appeared, and what logical groups they fell into for master keying purposes, but did not give us any new combinations.

This section deals with a slightly different METHOD of progressing the system, which typically gives us more combinations for change keys than does the TOTAL POSITION PROGRESSION or the PARTIAL PROGRESSION TECHNIQUE.

The disadvantage of it is that we must be far more careful in breaking the system up into smaller master key groups, because each of the master keys involved has far greater power than in the systems built by the other two methods. Also, in the other methods, each master had relatively unique operational control of its change keys. In the method in this chapter, there is a certain amount of overlapping that we must watch out for in selecting our lower levels of master keys.

This method is known in the trade as **ROTATING CONSTANT** (Older terms no longer deemed appropriate are "**HOLD AND VARY**" and "**MOVABLE CONSTANT**.)

It should be mentioned, however, that methods NOT related to this one have also been known by the same exact names, to reflect an individual's teaching STYLE for the Total Position Progression. Care should thus be taken when discussing a system to determine the exact meaning of terms such as these, to avoid confusion.

The one restriction of the Hold and Vary method is that it can NEVER be mixed with a Total Position Progression Method master key system. The rules with which we develop a Rotating Constant system will AUTOMATICALLY cause the change keys to interchange drastically (as though they were masters) with the change keys from the other method.

However, this only applies to true TPP or Total Position Progression. It does not apply to the LIMITED PROGRESSION technique, even if it were originally written as a Total Position Progression system and then broken into smaller master key systems using Limited Progression.

Once you fully understand the Total Position Progression Method and the Limited Progression Method, it is easy to understand and develop a Rotating Constant system. The rules are only slightly different, and the math is virtually the same.

Our first step, as in the other methods, is to fill out the Possible Depths section of a KBA Chart. Then we select our increment size, select a TMK and using the chart and guidelines fill out the bottom (cut possibilities section of the chart.

In the Total Position Progression method, our next step was to select a progression sequence and begin development of the system. In the Hold and Vary Method, however, we will be breaking it up into several DIFFERENT progression sequences instead.

One of the rules of the Total Position Progression Method (but not the Limited Progression Method) was that the same depth cut never appeared in the same position on both the change key and the TMK. In the Rotating Constant Method, the exact OPPOSITE is true.

**The "CONSTANT" refers to "holding constant" a CERTAIN NUMBER OF CUTS from the master key, and applying them to the change keys. Unlike the Limited Progression technique, however, it will not be the same cuts in the same positions on all the keys in the system.**

**BUT ALL THE CHANGE KEYS IN THE SYSTEM MUST HOLD THE SAME NUMBER OF CUTS FROM THE THEORETICAL MASTER!!!**

That is an ABSOLUTE rule for Rotating Constant.

## PROGRESSION PATTERNS FOR ROTATING CONSTANT SYSTEMS

As mentioned in the introduction to the Rotating Constant method, we will be "HOLDING CONSTANT" a certain number of cuts from the TMK on each key, although WHICH cuts were held will vary from key to key, as will which columns (or positions on the key) that we progress. In the Partial Progression Technique section we saw that the more positions from the TMK that were held, the fewer the number of possible change keys.

In order to understand this better, let's look at one group of progression patterns. Rotating Constant progression patterns are frequently identified by the numbers of the positions which we choose to hold. These positions are identified from bow to tip, as shown in this illustration:

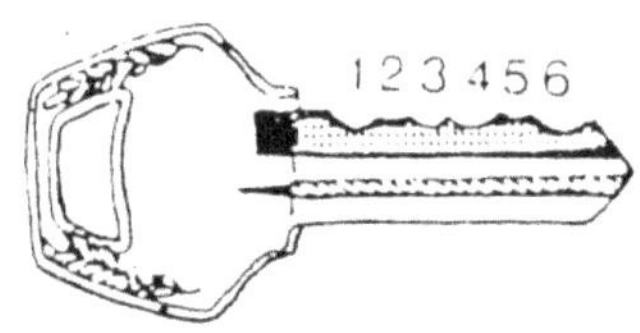


The first group we will look at consists of a six pin lock, where we have decided to hold five positions. For simplicity, we will assume that we have a manufacturer's numbering system of zero to nine to work with. Because we have six cuts on the key (for a six pin lock), and are "HOLDING CONSTANT" five positions, this means that we will be PROGRESSING one position. It's easy to see that we have six choices as to which position we will progress.

Let's assume a theoretical master of 238761. If we chose to progress only the sixth position, the base field would look like this:

| 2 | 3 | 8 | 7 | 6 | 1 |
|---|---|---|---|---|---|
| | | | | | 3 |
| | | | | | 5 |
| | | | | | 7 |
| | | | | | 9 |

and would give us the combinations:

2 3 8 7 6 3
2 3 8 7 6 5
2 3 8 7 6 7
2 3 8 7 6 9

with basic pinning of:

with basic pinning of:

| | | | |
|---|---|---|---|
| MK 238761 | MK 238761 | MK 238761 | MK 238761 |
| CK 238763 | CK 238765 | CK 238767 | CK 238769 |
| BP 238761 | BP 238761 | BP 238761 | BP 238761 |
| MP -----2 | MP -----4 | MP -----6 | MP -----8 |

If we chose to progress only the fifth position, the base field would look like this:

| 2 | 3 | 8 | 7 | 6 | 1 |
|---|---|---|---|---|---|
| | | | | 0 | |
| | | | | 2 | |
| | | | | 4 | |
| | | | | 8 | |

and would give us the combinations:

2 3 8 7 0 1
2 3 8 7 2 1
2 3 8 7 4 1
2 3 8 7 8 1

a) with basic pinning of:

| | | | |
|---|---|---|---|
| MK 238761 | MK 238761 | MK 238761 | MK 238761 |
| CK 238701 | CK 238721 | CK 238741 | CK 238781 |
| BP 238701 | BP 238721 | BP 238741 | BP 238761 |
| MP ----6- | MP ----4- | MP ----2- | MP ----2- |

If we chose to progress only the fourth position, the base field would look like this:

| 2 | 3 | 8 | 7 | 6 | 1 |
|---|---|---|---|---|---|
| | | | 1 | | |
| | | | 3 | | |
| | | | 5 | | |
| | | | 9 | | |

and would give us the combinations:

2 3 8 1 6 1
2 3 8 3 6 1
2 3 8 5 6 1
2 3 8 9 6 1

with basic pinning of:

| | | | |
|---|---|---|---|
| MK 238761 | MK 238761 | MK 238761 | MK 238761 |
| CK 238161 | CK 238361 | CK 238561 | CK 238961 |
| BP 238161 | BP 238361 | BP 238561 | BP 238761 |
| MP ---6-- | MP ---4-- | MP ---2-- | MP ---2-- |

If we chose to progress only the third position, the base field would look like this:

| 2 | 3 | 8 | 7 | 6 | 1 |
|---|---|---|---|---|---|
| | | 0 | | | |
| | | 2 | | | |
| | | 4 | | | |
| | | 6 | | | |

and would give us the combinations:

2 3 0 7 6 1
2 3 2 7 6 1
2 3 4 7 6 1
2 3 6 7 6 1

with basic pinning of:

| | | | |
|---|---|---|---|
| MK 238761 | MK 238761 | MK 238761 | MK 238761 |
| CK 230761 | CK 232761 | CK 234761 | CK 236761 |
| BP 230761 | BP 232761 | BP 234761 | BP 236761 |
| MP --8--- | MP --6--- | MP --4--- | MP --2--- |

If we chose to progress only the second position, the base field would look like this:

| 2 | 3 | 8 | 7 | 6 | 1 |
|---|---|---|---|---|---|
| | 1 | | | | |
| | 5 | | | | |
| | 7 | | | | |
| | 9 | | | | |

and would give us the combinations:

2 1 8 7 6 1
2 5 8 7 6 1
2 7 8 7 6 1
2 9 8 7 6 1

with basic pinning of:

| | | | |
|---|---|---|---|
| MK 238761 | MK 238761 | MK 238761 | MK 238761 |
| CK 218761 | CK 258761 | CK 278761 | CK 298761 |
| BP 218761 | BP 238761 | BP 238761 | BP 238761 |
| MP -2---- | MP -2---- | MP -4---- | MP -6---- |

If we chose to progress only the first position, the base field would look like this:

| 2 | 3 | 8 | 7 | 6 | 1 |
|---|---|---|---|---|---|
| 0 | | | | | |
| 4 | | | | | |
| 6 | | | | | |
| 8 | | | | | |

and would give us the combinations:

```
0 3 8 7 6 1
4 3 8 7 6 1
6 3 8 7 6 1
8 3 8 7 6 1
```

with basic pinning of:

```
MK 238761    MK 238761    MK 238761    MK 238761
CK 038761    CK 438761    CK 639761    CK 838761
BP 038761    BP 238761    BP 238761    BP 238761
MP 2-----    MP 2-----    MP 4-----    MP 6-----
```

Notice that in each of those six possible systems that we developed, the theoretical master controlled 4 combinations, each of which, when the lock was pinned, would only have one master pin in it.

BUT WAIT A MINUTE! The master didn't change a bit, and none of the change keys we developed would interchange with any of the others, even though the master would operate all of them!

Congratulations! You've just seen a Rotating Constant system developed from scratch. This would be a six pin system HOLDING FIVE POSITIONS CONSTANT from the TMK. There are actually six progression patterns involved:

**12345 12346 12356 12456 13456 23456**

(The patterns are identified by the positions where we "HOLD" a master cut.) which could also be written as:

CCCCC**P** CCCC**P**C CCC**P**CC CC**P**CCC C**P**CCCC **P**CCCCC

Where C indicates a position we have held the master cut CONSTANT for, and **P** indicates a position where we will PROGRESS the cuts, the following is true:

Pattern 12345 would be CCCCC**P**
Pattern 12346 would be CCCC**P**C
Pattern 12356 would be CCC**P**CC

Pattern 12456 would be CC**P**CCC
Pattern 13456 would be C**P**CCCC
Pattern 23456 would be **P**CCCCC

This method gave us 24 combinations for change keys under one master, with only one master pin per lock! In the Total Position Progression method or using the Partial Progression Technique, on the same brand of locks, with the same master key pattern, one master pin per lock would have given us only 4 possible change keys. Two master pins per lock would have given us only 16 possible change keys. In fact, to control 24 possible change keys, we would have had to put THREE master pins in each and every lock in the system! As we saw earlier in the book, the more master pins in the lock, the less security the lock has. The less master pins, the better security the lock has. So the Rotating Constant Method, on a simple system like this, increases both the number of possible change keys, and the security of the lock!

The next two pages show the progression patterns for a six pin lock cylinder and for a five pin lock cylinder.

The number of change keys that we end up with under Rotating Constant will depend on the numbering system the lock manufacturer uses, whether the locks are 5 pin or 6 pin, how many positions we "HOLD CONSTANT" from the TMK, and the options we chose earlier.

**REMEMBER THAT WE CANNOT COMBINE PATTERNS THAT HOLD A DIFFERENT NUMBER OF POSITIONS, INTO THE SAME SYSTEM. WE MUST EITHER "HOLD ONE" OR "HOLD TWO" OR ETC.**

There are two variants on the RCM technique that do not break this rule but at first glance may seem to. We will cover them in later lessons. They are:

The XCM Technique of Rotating Constant Method
and
The RCM with Selective Master Keys

With both of these techniques, a level of planning and understanding greater than simple RCM will be required, but the rewards make them unbelievably effective techniques, and very difficult to "crack", "decode" or "key into".

But those same strengths could make them dangerous when attempted by those not fully skilled and knowledgeable in the Rotating Constant Method (RCM) so if you want to add them to your master keying toolkit, you will need to pay close attention to the this chapter and the next chapter, and be certain you fully understand the basic RCM layout and implementation.

**The progression patterns for a six pin lock would be:**

**HOLDING ONE (POSITION) (5 master pins per lock)**

1 CPPPPP
2 PCPPPP
3 PPCPPP
4 PPPCPP
5 PPPPCP
6 PPPPPC

**HOLDING TWO (POSITIONS) (4 master pins per lock)**

12 CCPPPP
13 CPCPPP 23 PCCPPP
14 CPPCPP 24 PCPCPP 34 PPCCPP
15 CPPPCP 25 PCPPCP 35 PPCPCP 45 PPPCCP
16 CPPPPC 26 PCPPPC 36 PPCPPC 46 PPPCPC 56 PPPPCC

**HOLDING THREE (POSITIONS) (3 master pins per lock)**

123 CCCPPP
124 CCPCPP
125 CCPPCP
126 CCPPPC
134 CPCCPP 234 PCCCPP
135 CPCPCP 235 PCCPCP
136 CPCPPC 236 PCCPPC
145 CPPCCP 245 PCPCCP 345 PPCCCP
146 CPPCPC 246 PCPCPC 346 PPCCPC
156 CPPPCC 256 PCPPCC 356 PPCPCC 456 PPPCCC

**HOLDING FOUR (POSITIONS) (2 master pins per lock)**

1234 CCCCPP
1235 CCCPCP
1236 CCCPPC
1245 CCPCCP
1246 CCPCPC
1256 CCPPCC
1345 CPCCCP 2345 PCCCCP
1346 CPCCPC 2346 PCCCPC
1356 CPCPCC 2356 PCCPCC
1456 CPPCCC 2456 PCPCCC 3456 PPCCCC

**HOLDING FIVE (POSITIONS) (1 master pin per lock)**

12345 CCCCCP 12456 CCPCCC 12346 CCCCPC
13456 CPCCCC 12356 CCCPCC 23456 PCCCCC

**The progression patterns for a five pin lock would be:**

**HOLDING ONE (POSITION) (4 master pins per lock)**
1 C**PPPP**
2 **P**C**PPP**
3 **PP**C**PP**
4 **PPP**C**P**
5 **PPPP**C

**HOLDING TWO (POSITIONS) (3 master pins per lock)**
12 CC**PPP**
13 C**P**C**PP** 23 **P**CC**PP**
14 C**PP**C**P** 24 **P**C**P**C**P** 34 **PP**CC**P**
15 C**PPP**C 25 **P**C**PP**C 35 **PP**C**P**C 45 **PPP**CC

**HOLDING THREE (POSITIONS) (2 master pins per lock)**
123 CCC**PP**
124 CC**P**C**P**
125 CC**PP**C
134 C**P**CC**P** 234 **P**CCC**P**
135 C**P**C**P**C 235 **P**CC**P**C
145 C**PP**CC 245 **P**C**P**CC 345 **PP**CCC

**HOLDING FOUR (POSITIONS) (1 master pin per lock)**
1234 CCCC**P**
1235 CCC**P**C
1245 CC**P**CC
1345 C**P**CCC 2345 **P**CCCC

Now let's look at a sample system worked out for Holding 4 positions Constant of 6. We can use the standard charts as described on the previous page.

**Note that each progressed position is progressed, not in terms of its position on the key, but in terms of its progression sequence in that particular Rotating Constant Pattern.**

Bear in mind that in this pattern, there are not any Horizontal Masters, Line Masters, or Page Masters. The highest key in this system is the TMK. Each of the keys in the system has only two cuts different from it. Therefore, it could be said in this particular example, that the TMK is a Vertical Master, although as we see, it operates multiple Vertical columns (one per pattern on pattern) instead of the more traditional one column from TPP or using the Partial Progression technique.

Therefore, there can be no additional Level 2 masters. The Block Master capability is still present, though, as you will see shortly, but with changed capabilities.

The progression would have been the same if we had used standard charts, as long as we keep in mind which progression pattern we are dealing with. It usually helps to write it above the particular chart you are using. Remember that you get a new First Progression Sequence, Second Progression Sequence, etc with each chart. They will not be in the same positions from chart to chart.

Remember that if the pattern is "Progressing 1", each progression pattern will require only a Block to progress. If it is "Progressing 2", as in this example, it will require a Vertical Column for each pattern. If it is "Progressing 3", it will require a complete page for each progression pattern, etc.

Let's look at a RCM Pattern of Holding 4 Constant and Progressing 2 on a six pin (also known as Holding 4 of 6) and see how the 15 patterns develop.

**Remember that although we are using a standard SPF chart to write them, we are actually writing 4 separate charts (progression patterns) on each sheet.**

**There is no direct relationship between the patterns shown on the chart. We could have done each pattern on a separate page of the chart. It would simply have taken up more space and more paper.**

**We have no Line Master Keys or Horizontal Master Keys under this pattern because there are no lines or Horizontal columns. That does not mean there will not be a "c parallel master" or an "ac parallel master" with some authority within the system. In fact, these may become change keys under some patterns!**

**Each pattern is simply a Vertical Column.**

5 4 3 2 5 8

VERTICAL MASTER

| C | C | C | C | P | P |
|---|---|---|---|---|---|
| 5 | 4 | 3 | 2 | 7 | 0 |
| 5 | 4 | 3 | 2 | 7 | 2 |
| 5 | 4 | 3 | 2 | 7 | 4 |
| 5 | 4 | 3 | 2 | 7 | 6 |

BLOCK MASTER

| C | C | C | P | C | P |
|---|---|---|---|---|---|
| 5 | 4 | 3 | 4 | 5 | 0 |
| 5 | 4 | 3 | 4 | 5 | 2 |
| 5 | 4 | 3 | 4 | 5 | 4 |
| 5 | 4 | 3 | 4 | 5 | 6 |

BLOCK MASTER

| C | C | P | C | C | P |
|---|---|---|---|---|---|
| 5 | 4 | 5 | 2 | 5 | 0 |
| 5 | 4 | 5 | 2 | 5 | 2 |
| 5 | 4 | 5 | 2 | 5 | 4 |
| 5 | 4 | 5 | 2 | 5 | 6 |

BLOCK MASTER

| C | P | C | C | C | P |
|---|---|---|---|---|---|
| 5 | 6 | 3 | 2 | 5 | 0 |
| 5 | 6 | 3 | 2 | 5 | 2 |
| 5 | 6 | 3 | 2 | 5 | 4 |
| 5 | 6 | 3 | 2 | 5 | 6 |

BLOCK MASTER

| | | | | | |
|---|---|---|---|---|---|
| 5 | 4 | 3 | 2 | 9 | 0 |
| 5 | 4 | 3 | 2 | 9 | 2 |
| 5 | 4 | 3 | 2 | 9 | 4 |
| 5 | 4 | 3 | 2 | 9 | 6 |

BLOCK MASTER

| | | | | | |
|---|---|---|---|---|---|
| 5 | 4 | 3 | 6 | 5 | 0 |
| 5 | 4 | 3 | 6 | 5 | 2 |
| 5 | 4 | 3 | 6 | 5 | 4 |
| 5 | 4 | 3 | 6 | 5 | 6 |

BLOCK MASTER

| | | | | | |
|---|---|---|---|---|---|
| 5 | 4 | 7 | 2 | 5 | 0 |
| 5 | 4 | 7 | 2 | 5 | 2 |
| 5 | 4 | 7 | 2 | 5 | 4 |
| 5 | 4 | 7 | 2 | 5 | 6 |

BLOCK MASTER

| | | | | | |
|---|---|---|---|---|---|
| 5 | 8 | 3 | 2 | 5 | 0 |
| 5 | 8 | 3 | 2 | 5 | 2 |
| 5 | 8 | 3 | 2 | 5 | 4 |
| 5 | 8 | 3 | 2 | 5 | 6 |

BLOCK MASTER

| | | | | | |
|---|---|---|---|---|---|
| 5 | 4 | 3 | 2 | 1 | 0 |
| 5 | 4 | 3 | 2 | 1 | 2 |
| 5 | 4 | 3 | 2 | 1 | 4 |
| 5 | 4 | 3 | 2 | 1 | 6 |

BLOCK MASTER

| | | | | | |
|---|---|---|---|---|---|
| 5 | 4 | 3 | 8 | 5 | 0 |
| 5 | 4 | 3 | 8 | 5 | 2 |
| 5 | 4 | 3 | 8 | 5 | 4 |
| 5 | 4 | 3 | 8 | 5 | 6 |

BLOCK MASTER

| | | | | | |
|---|---|---|---|---|---|
| 5 | 4 | 9 | 2 | 5 | 0 |
| 5 | 4 | 9 | 2 | 5 | 2 |
| 5 | 4 | 9 | 2 | 5 | 4 |
| 5 | 4 | 9 | 2 | 5 | 6 |

BLOCK MASTER

| | | | | | |
|---|---|---|---|---|---|
| 5 | 0 | 3 | 2 | 5 | 0 |
| 5 | 0 | 3 | 2 | 5 | 2 |
| 5 | 0 | 3 | 2 | 5 | 4 |
| 5 | 0 | 3 | 2 | 5 | 6 |

BLOCK MASTER

| | | | | | |
|---|---|---|---|---|---|
| 5 | 4 | 3 | 2 | 3 | 0 |
| 5 | 4 | 3 | 2 | 3 | 2 |
| 5 | 4 | 3 | 2 | 3 | 4 |
| 5 | 4 | 3 | 2 | 3 | 6 |

BLOCK MASTER

| | | | | | |
|---|---|---|---|---|---|
| 5 | 4 | 3 | 0 | 5 | 0 |
| 5 | 4 | 3 | 0 | 5 | 2 |
| 5 | 4 | 3 | 0 | 5 | 4 |
| 5 | 4 | 3 | 0 | 5 | 6 |

BLOCK MASTER

| | | | | | |
|---|---|---|---|---|---|
| 5 | 4 | 1 | 2 | 5 | 0 |
| 5 | 4 | 1 | 2 | 5 | 2 |
| 5 | 4 | 1 | 2 | 5 | 4 |
| 5 | 4 | 1 | 2 | 5 | 6 |

BLOCK MASTER

| | | | | | |
|---|---|---|---|---|---|
| 5 | 2 | 3 | 2 | 5 | 0 |
| 5 | 2 | 3 | 2 | 5 | 2 |
| 5 | 2 | 3 | 2 | 5 | 4 |
| 5 | 2 | 3 | 2 | 5 | 6 |

BLOCK MASTER

| P | C | C | C | C | P |
|---|---|---|---|---|---|
| 7 | 4 | 3 | 2 | 5 | 0 |
| 7 | 4 | 3 | 2 | 5 | 2 |
| 7 | 4 | 3 | 2 | 5 | 4 |
| 7 | 4 | 3 | 2 | 5 | 6 |

BLOCK MASTER

| C | C | C | P | P | C |
|---|---|---|---|---|---|
| 5 | 4 | 3 | 4 | 7 | 8 |
| 5 | 4 | 3 | 4 | 9 | 8 |
| 5 | 4 | 3 | 4 | 1 | 8 |
| 5 | 4 | 3 | 4 | 3 | 8 |

BLOCK MASTER

| C | C | P | C | P | C |
|---|---|---|---|---|---|
| 5 | 4 | 5 | 2 | 7 | 8 |
| 5 | 4 | 5 | 2 | 9 | 8 |
| 5 | 4 | 5 | 2 | 1 | 8 |
| 5 | 4 | 5 | 2 | 3 | 8 |

BLOCK MASTER

| C | P | C | C | P | C |
|---|---|---|---|---|---|
| 5 | 6 | 3 | 2 | 7 | 8 |
| 5 | 6 | 3 | 2 | 9 | 8 |
| 5 | 6 | 3 | 2 | 1 | 8 |
| 5 | 6 | 3 | 2 | 3 | 8 |

BLOCK MASTER

| 9 | 4 | 3 | 2 | 5 | 0 |
|---|---|---|---|---|---|
| 9 | 4 | 3 | 2 | 5 | 2 |
| 9 | 4 | 3 | 2 | 5 | 4 |
| 9 | 4 | 3 | 2 | 5 | 6 |

BLOCK MASTER

| 5 | 4 | 3 | 6 | 7 | 8 |
|---|---|---|---|---|---|
| 5 | 4 | 3 | 6 | 9 | 8 |
| 5 | 4 | 3 | 6 | 1 | 8 |
| 5 | 4 | 3 | 6 | 3 | 8 |

BLOCK MASTER

| 5 | 4 | 7 | 2 | 7 | 8 |
|---|---|---|---|---|---|
| 5 | 4 | 7 | 2 | 9 | 8 |
| 5 | 4 | 7 | 2 | 1 | 8 |
| 5 | 4 | 7 | 2 | 3 | 8 |

BLOCK MASTER

| 5 | 8 | 3 | 2 | 7 | 8 |
|---|---|---|---|---|---|
| 5 | 8 | 3 | 2 | 9 | 8 |
| 5 | 8 | 3 | 2 | 1 | 8 |
| 5 | 8 | 3 | 2 | 3 | 8 |

BLOCK MASTER

| 1 | 4 | 3 | 2 | 5 | 0 |
|---|---|---|---|---|---|
| 1 | 4 | 3 | 2 | 5 | 2 |
| 1 | 4 | 3 | 2 | 5 | 4 |
| 1 | 4 | 3 | 2 | 5 | 6 |

BLOCK MASTER

| 5 | 4 | 3 | 8 | 7 | 8 |
|---|---|---|---|---|---|
| 5 | 4 | 3 | 8 | 9 | 8 |
| 5 | 4 | 3 | 8 | 1 | 8 |
| 5 | 4 | 3 | 8 | 3 | 8 |

BLOCK MASTER

| 5 | 4 | 9 | 2 | 7 | 8 |
|---|---|---|---|---|---|
| 5 | 4 | 9 | 2 | 9 | 8 |
| 5 | 4 | 9 | 2 | 1 | 8 |
| 5 | 4 | 9 | 2 | 3 | 8 |

BLOCK MASTER

| 5 | 0 | 3 | 2 | 7 | 8 |
|---|---|---|---|---|---|
| 5 | 0 | 3 | 2 | 9 | 8 |
| 5 | 0 | 3 | 2 | 1 | 8 |
| 5 | 0 | 3 | 2 | 3 | 8 |

BLOCK MASTER

| 3 | 4 | 3 | 2 | 5 | 0 |
|---|---|---|---|---|---|
| 3 | 4 | 3 | 2 | 5 | 2 |
| 3 | 4 | 3 | 2 | 5 | 4 |
| 3 | 4 | 3 | 2 | 5 | 6 |

BLOCK MASTER

| 5 | 4 | 3 | 0 | 7 | 8 |
|---|---|---|---|---|---|
| 5 | 4 | 3 | 0 | 9 | 8 |
| 5 | 4 | 3 | 0 | 1 | 8 |
| 5 | 4 | 3 | 0 | 3 | 8 |

BLOCK MASTER

| 5 | 4 | 1 | 2 | 7 | 8 |
|---|---|---|---|---|---|
| 5 | 4 | 1 | 2 | 9 | 8 |
| 5 | 4 | 1 | 2 | 1 | 8 |
| 5 | 4 | 1 | 2 | 3 | 8 |

BLOCK MASTER

| 5 | 2 | 3 | 2 | 7 | 8 |
|---|---|---|---|---|---|
| 5 | 2 | 3 | 2 | 9 | 8 |
| 5 | 2 | 3 | 2 | 1 | 8 |
| 5 | 2 | 3 | 2 | 3 | 8 |

BLOCK MASTER

| P | C | C | C | P | C |
|---|---|---|---|---|---|
| 7 | 4 | 3 | 2 | 7 | 8 |
| 7 | 4 | 3 | 2 | 9 | 8 |
| 7 | 4 | 3 | 2 | 1 | 8 |
| 7 | 4 | 3 | 2 | 3 | 8 |

BLOCK MASTER

| C | C | P | P | C | C |
|---|---|---|---|---|---|
| 5 | 4 | 5 | 4 | 5 | 8 |
| 5 | 4 | 5 | 6 | 5 | 8 |
| 5 | 4 | 5 | 8 | 5 | 8 |
| 5 | 4 | 5 | 0 | 5 | 8 |

BLOCK MASTER

| C | P | C | P | C | C |
|---|---|---|---|---|---|
| 5 | 6 | 3 | 4 | 5 | 8 |
| 5 | 6 | 3 | 6 | 5 | 8 |
| 5 | 6 | 3 | 8 | 5 | 8 |
| 5 | 6 | 3 | 0 | 5 | 8 |

BLOCK MASTER

| P | C | C | P | C | C |
|---|---|---|---|---|---|
| 7 | 4 | 3 | 4 | 5 | 8 |
| 7 | 4 | 3 | 6 | 5 | 8 |
| 7 | 4 | 3 | 8 | 5 | 8 |
| 7 | 4 | 3 | 0 | 5 | 8 |

BLOCK MASTER

| 9 | 4 | 3 | 2 | 7 | 8 |
|---|---|---|---|---|---|
| 9 | 4 | 3 | 2 | 9 | 8 |
| 9 | 4 | 3 | 2 | 1 | 8 |
| 9 | 4 | 3 | 2 | 3 | 8 |

BLOCK MASTER

| 5 | 4 | 7 | 4 | 5 | 8 |
|---|---|---|---|---|---|
| 5 | 4 | 7 | 6 | 5 | 8 |
| 5 | 4 | 7 | 8 | 5 | 8 |
| 5 | 4 | 7 | 0 | 5 | 8 |

BLOCK MASTER

| 5 | 8 | 3 | 4 | 5 | 8 |
|---|---|---|---|---|---|
| 5 | 8 | 3 | 6 | 5 | 8 |
| 5 | 8 | 3 | 8 | 5 | 8 |
| 5 | 8 | 3 | 0 | 5 | 8 |

BLOCK MASTER

| 9 | 4 | 3 | 4 | 5 | 8 |
|---|---|---|---|---|---|
| 9 | 4 | 3 | 6 | 5 | 8 |
| 9 | 4 | 3 | 8 | 5 | 8 |
| 9 | 4 | 3 | 0 | 5 | 8 |

BLOCK MASTER

| 1 | 4 | 3 | 2 | 7 | 8 |
|---|---|---|---|---|---|
| 1 | 4 | 3 | 2 | 9 | 8 |
| 1 | 4 | 3 | 2 | 1 | 8 |
| 1 | 4 | 3 | 2 | 3 | 8 |

BLOCK MASTER

| 5 | 4 | 9 | 4 | 5 | 8 |
|---|---|---|---|---|---|
| 5 | 4 | 9 | 6 | 5 | 8 |
| 5 | 4 | 9 | 8 | 5 | 8 |
| 5 | 4 | 9 | 0 | 5 | 8 |

BLOCK MASTER

| 5 | 0 | 3 | 4 | 5 | 8 |
|---|---|---|---|---|---|
| 5 | 0 | 3 | 6 | 5 | 8 |
| 5 | 0 | 3 | 8 | 5 | 8 |
| 5 | 0 | 3 | 0 | 5 | 8 |

BLOCK MASTER

| 1 | 4 | 3 | 4 | 5 | 8 |
|---|---|---|---|---|---|
| 1 | 4 | 3 | 6 | 5 | 8 |
| 1 | 4 | 3 | 8 | 5 | 8 |
| 1 | 4 | 3 | 0 | 5 | 8 |

BLOCK MASTER

| 3 | 4 | 3 | 2 | 7 | 8 |
|---|---|---|---|---|---|
| 3 | 4 | 3 | 2 | 9 | 8 |
| 3 | 4 | 3 | 2 | 1 | 8 |
| 3 | 4 | 3 | 2 | 3 | 8 |

BLOCK MASTER

| 5 | 4 | 1 | 4 | 5 | 8 |
|---|---|---|---|---|---|
| 5 | 4 | 1 | 6 | 5 | 8 |
| 5 | 4 | 1 | 8 | 5 | 8 |
| 5 | 4 | 1 | 0 | 5 | 8 |

BLOCK MASTER

| 5 | 2 | 3 | 4 | 5 | 8 |
|---|---|---|---|---|---|
| 5 | 2 | 3 | 6 | 5 | 8 |
| 5 | 2 | 3 | 8 | 5 | 8 |
| 5 | 2 | 3 | 0 | 5 | 8 |

BLOCK MASTER

| 3 | 4 | 3 | 4 | 5 | 8 |
|---|---|---|---|---|---|
| 3 | 4 | 3 | 6 | 5 | 8 |
| 3 | 4 | 3 | 8 | 5 | 8 |
| 3 | 4 | 3 | 0 | 5 | 8 |

BLOCK MASTER

| P | C | P | C | C | C |
|---|---|---|---|---|---|
| 7 | 4 | 5 | 2 | 5 | 8 |
| 7 | 4 | 7 | 2 | 5 | 8 |
| 7 | 4 | 9 | 2 | 5 | 8 |
| 7 | 4 | 1 | 2 | 5 | 8 |

BLOCK MASTER

| C | P | P | C | C | C |
|---|---|---|---|---|---|
| 5 | 6 | 5 | 2 | 5 | 8 |
| 5 | 6 | 7 | 2 | 5 | 8 |
| 5 | 6 | 9 | 2 | 5 | 8 |
| 5 | 6 | 1 | 2 | 5 | 8 |

BLOCK MASTER

| P | P | C | C | C | C |
|---|---|---|---|---|---|
| 7 | 6 | 3 | 2 | 5 | 8 |
| 7 | 8 | 3 | 2 | 5 | 8 |
| 7 | 0 | 3 | 2 | 5 | 8 |
| 7 | 2 | 3 | 2 | 5 | 8 |

BLOCK MASTER

| | | | | | |
|---|---|---|---|---|---|
| | | | | | |
| | | | | | |
| | | | | | |
| | | | | | |

BLOCK MASTER

| | | | | | |
|---|---|---|---|---|---|
| 9 | 4 | 5 | 2 | 5 | 8 |
| 9 | 4 | 7 | 2 | 5 | 8 |
| 9 | 4 | 9 | 2 | 5 | 8 |
| 9 | 4 | 1 | 2 | 5 | 8 |

BLOCK MASTER

| | | | | | |
|---|---|---|---|---|---|
| 5 | 8 | 5 | 2 | 5 | 8 |
| 5 | 8 | 7 | 2 | 5 | 8 |
| 5 | 8 | 9 | 2 | 5 | 8 |
| 5 | 8 | 1 | 2 | 5 | 8 |

BLOCK MASTER

| | | | | | |
|---|---|---|---|---|---|
| 9 | 6 | 3 | 2 | 5 | 8 |
| 9 | 8 | 3 | 2 | 5 | 8 |
| 9 | 0 | 3 | 2 | 5 | 8 |
| 9 | 2 | 3 | 2 | 5 | 8 |

BLOCK MASTER

| | | | | | |
|---|---|---|---|---|---|
| | | | | | |
| | | | | | |
| | | | | | |
| | | | | | |

BLOCK MASTER

| | | | | | |
|---|---|---|---|---|---|
| 1 | 4 | 5 | 2 | 5 | 8 |
| 1 | 4 | 7 | 2 | 5 | 8 |
| 1 | 4 | 9 | 2 | 5 | 8 |
| 1 | 4 | 1 | 2 | 5 | 8 |

BLOCK MASTER

| | | | | | |
|---|---|---|---|---|---|
| 5 | 0 | 5 | 2 | 5 | 8 |
| 5 | 0 | 7 | 2 | 5 | 8 |
| 5 | 0 | 9 | 2 | 5 | 8 |
| 5 | 0 | 1 | 2 | 5 | 8 |

BLOCK MASTER

| | | | | | |
|---|---|---|---|---|---|
| 1 | 6 | 3 | 2 | 5 | 8 |
| 1 | 8 | 3 | 2 | 5 | 8 |
| 1 | 0 | 3 | 2 | 5 | 8 |
| 1 | 2 | 3 | 2 | 5 | 8 |

BLOCK MASTER

| | | | | | |
|---|---|---|---|---|---|
| | | | | | |
| | | | | | |
| | | | | | |
| | | | | | |

BLOCK MASTER

| | | | | | |
|---|---|---|---|---|---|
| 3 | 4 | 5 | 2 | 5 | 8 |
| 3 | 4 | 7 | 2 | 5 | 8 |
| 3 | 4 | 9 | 2 | 5 | 8 |
| 3 | 4 | 1 | 2 | 5 | 8 |

BLOCK MASTER

| | | | | | |
|---|---|---|---|---|---|
| 5 | 2 | 5 | 2 | 5 | 8 |
| 5 | 2 | 7 | 2 | 5 | 8 |
| 5 | 2 | 9 | 2 | 5 | 8 |
| 5 | 2 | 1 | 2 | 5 | 8 |

BLOCK MASTER

| | | | | | |
|---|---|---|---|---|---|
| 3 | 6 | 3 | 2 | 5 | 8 |
| 3 | 8 | 3 | 2 | 5 | 8 |
| 3 | 0 | 3 | 2 | 5 | 8 |
| 3 | 2 | 3 | 2 | 5 | 8 |

BLOCK MASTER

| | | | | | |
|---|---|---|---|---|---|
| | | | | | |
| | | | | | |
| | | | | | |
| | | | | | |

BLOCK MASTER

**Next, let's look at one of the more common RCM patterns, holding three constant out of six positions. In this case, each of the 20 progression patterns will use an entire SPF chart to develop its bitting combinations.**

| KBA | | | | | | |
|---|---|---|---|---|---|---|
| 5 | 4 | 3 | 0 | 5 | 8 | THEORETICAL MASTER |
| 3 | 2 | 1 | 2 | 7 | 0 | BASE FIELD (KBA) KEY BITTING ARRAY |
| 7 | 6 | 5 | 4 | 9 | 2 | |
| 9 | 8 | 7 | 6 | 1 | 4 | |
| 1 | 0 | 9 | 8 | 3 | 6 | |
| C | C | C | P | P | P | ⇦ SEQUENCE OF PROGRESSION |

PAGE MASTER ______________ PAGE # ________

GROUP MASTER ______________ OF ________ PAGES

SECTION MASTER ______________

| | # | VERTICAL MASTER | # | VERTICAL MASTER | # | VERTICAL MASTER | # | VERTICAL MASTER | LINE MASTERS ⇩ |
|---|---|---|---|---|---|---|---|---|---|
| | 1 | 5 4 3 2 7 0 | 17 | 5 4 3 4 7 0 | 33 | 5 4 3 6 7 0 | 49 | 5 4 3 8 7 0 | |
| | 2 | 5 4 3 2 7 2 | 18 | 5 4 3 4 7 2 | 34 | 5 4 3 6 7 2 | 50 | 5 4 3 8 7 2 | |
| HORIZONTAL MASTER | 3 | 5 4 3 2 7 4 | 19 | 5 4 3 4 7 4 | 35 | 5 4 3 6 7 4 | 51 | 5 4 3 8 7 4 | |
| | 4 | 5 4 3 2 7 6 | 20 | 5 4 3 4 7 6 | 36 | 5 4 3 6 7 6 | 52 | 5 4 3 8 7 6 | |
| | | BLOCK MASTER | | BLOCK MASTER | | BLOCK MASTER | | BLOCK MASTER | |
| | 5 | 5 4 3 2 9 0 | 21 | 5 4 3 4 9 0 | 37 | 5 4 3 6 9 0 | 53 | 5 4 3 8 9 0 | |
| | 6 | 5 4 3 2 9 2 | 22 | 5 4 3 4 9 2 | 38 | 5 4 3 6 9 2 | 54 | 5 4 3 8 9 2 | |
| HORIZONTAL MASTER | 7 | 5 4 3 2 9 4 | 23 | 5 4 3 4 9 4 | 39 | 5 4 3 6 9 4 | 55 | 5 4 3 8 9 4 | |
| | 8 | 5 4 3 2 9 6 | 24 | 5 4 3 4 9 6 | 40 | 5 4 3 6 9 6 | 56 | 5 4 3 8 9 6 | |
| | | BLOCK MASTER | | BLOCK MASTER | | BLOCK MASTER | | BLOCK MASTER | |
| | 9 | 5 4 3 2 1 0 | 25 | 5 4 3 4 1 0 | 41 | 5 4 3 6 1 0 | 57 | 5 4 3 8 1 0 | |
| | 10 | 5 4 3 2 1 2 | 26 | 5 4 3 4 1 2 | 42 | 5 4 3 6 1 2 | 58 | 5 4 3 8 1 2 | |
| HORIZONTAL MASTER | 11 | 5 4 3 2 1 4 | 27 | 5 4 3 4 1 4 | 43 | 5 4 3 6 1 4 | 59 | 5 4 3 8 1 4 | |
| | 12 | 5 4 3 2 1 6 | 28 | 5 4 3 4 1 6 | 44 | 5 4 3 6 1 6 | 60 | 5 4 3 8 1 6 | |
| | | BLOCK MASTER | | BLOCK MASTER | | BLOCK MASTER | | BLOCK MASTER | |
| | 13 | 5 4 3 2 3 0 | 29 | 5 4 3 4 3 0 | 45 | 5 4 3 6 3 0 | 61 | 5 4 3 8 3 0 | |
| | 14 | 5 4 3 2 3 2 | 30 | 5 4 3 4 3 2 | 46 | 5 4 3 6 3 2 | 62 | 5 4 3 8 3 2 | |
| HORIZONTAL MASTER | 15 | 5 4 3 2 3 4 | 31 | 5 4 3 4 3 4 | 47 | 5 4 3 6 3 4 | 63 | 5 4 3 8 3 4 | |
| | 16 | 5 4 3 2 3 6 | 32 | 5 4 3 4 3 6 | 48 | 5 4 3 6 3 6 | 64 | 5 4 3 8 3 6 | |

| KBA | | | | | | |
|---|---|---|---|---|---|---|
| 5 | 4 | 3 | 0 | 5 | 8 | THEORETICAL MASTER |
| 3 | 2 | 1 | 2 | 7 | 0 | BASE FIELD (KBA) KEY BITTING ARRAY |
| 7 | 6 | 5 | 4 | 9 | 2 | |
| 9 | 8 | 7 | 6 | 1 | 4 | |
| 1 | 0 | 9 | 8 | 3 | 6 | |
| C | C | P | C | P | P | ⇦ SEQUENCE OF PROGRESSION |

| | # | VERTICAL MASTER | # | VERTICAL MASTER | # | VERTICAL MASTER | # | VERTICAL MASTER | LINE MASTERS ⇩ |
|---|---|---|---|---|---|---|---|---|---|
| | 1 | 5 4 1 0 7 0 | 17 | 5 4 5 0 7 0 | 33 | 5 4 7 0 7 0 | 49 | 5 4 9 0 7 0 | |
| | 2 | 5 4 1 0 7 2 | 18 | 5 4 5 0 7 2 | 34 | 5 4 7 0 7 2 | 50 | 5 4 9 0 7 2 | |
| HORIZONTAL MASTER | 3 | 5 4 1 0 7 4 | 19 | 5 4 5 0 7 4 | 35 | 5 4 7 0 7 4 | 51 | 5 4 9 0 7 4 | |
| | 4 | 5 4 1 0 7 6 | 20 | 5 4 5 0 7 6 | 36 | 5 4 7 0 7 6 | 52 | 5 4 9 0 7 6 | |
| | | BLOCK MASTER | | BLOCK MASTER | | BLOCK MASTER | | BLOCK MASTER | |
| | 5 | 5 4 1 0 9 0 | 21 | 5 4 5 0 9 0 | 37 | 5 4 7 0 9 0 | 53 | 5 4 9 0 9 0 | |
| | 6 | 5 4 1 0 9 2 | 22 | 5 4 5 0 9 2 | 38 | 5 4 7 0 9 2 | 54 | 5 4 9 0 9 2 | |
| HORIZONTAL MASTER | 7 | 5 4 1 0 9 4 | 23 | 5 4 5 0 9 4 | 39 | 5 4 7 0 9 4 | 55 | 5 4 9 0 9 4 | |
| | 8 | 5 4 1 0 9 6 | 24 | 5 4 5 0 9 6 | 40 | 5 4 7 0 9 6 | 56 | 5 4 9 0 9 6 | |
| | | BLOCK MASTER | | BLOCK MASTER | | BLOCK MASTER | | BLOCK MASTER | |
| | 9 | 5 4 1 0 1 0 | 25 | 5 4 5 0 1 0 | 41 | 5 4 7 0 1 0 | 57 | 5 4 9 0 1 0 | |
| | 10 | 5 4 1 0 1 2 | 26 | 5 4 5 0 1 2 | 42 | 5 4 7 0 1 2 | 58 | 5 4 9 0 1 2 | |
| HORIZONTAL MASTER | 11 | 5 4 1 0 1 4 | 27 | 5 4 5 0 1 4 | 43 | 5 4 7 0 1 4 | 59 | 5 4 9 0 1 4 | |
| | 12 | 5 4 1 0 1 6 | 28 | 5 4 5 0 1 6 | 44 | 5 4 7 0 1 6 | 60 | 5 4 9 0 1 6 | |
| | | BLOCK MASTER | | BLOCK MASTER | | BLOCK MASTER | | BLOCK MASTER | |
| | 13 | 5 4 1 0 3 0 | 29 | 5 4 5 0 3 0 | 45 | 5 4 7 0 3 0 | 61 | 5 4 9 0 3 0 | |
| | 14 | 5 4 1 0 3 2 | 30 | 5 4 5 0 3 2 | 46 | 5 4 7 0 3 2 | 62 | 5 4 9 0 3 2 | |
| HORIZONTAL MASTER | 15 | 5 4 1 0 3 4 | 31 | 5 4 5 0 3 4 | 47 | 5 4 7 0 3 4 | 63 | 5 4 9 0 3 4 | |
| | 16 | 5 4 1 0 3 6 | 32 | 5 4 5 0 3 6 | 48 | 5 4 7 0 3 6 | 64 | 5 4 9 0 3 6 | |
| | | BLOCK MASTER | | BLOCK MASTER | | BLOCK MASTER | | BLOCK MASTER | |

| 5 | 4 | 3 | 0 | 5 | 8 | THEORETICAL MASTER |
|---|---|---|---|---|---|---|
| 3 | 2 | 1 | 2 | 7 | 0 | BASE FIELD (KBA) KEY BITTING ARRAY |
| 7 | 6 | 5 | 4 | 9 | 2 | |
| 9 | 8 | 7 | 6 | 1 | 4 | |
| 1 | 0 | 9 | 8 | 3 | 6 | |
| C | C | P | P | C | P | ⇦ SEQUENCE OF PROGRESSION |

VERTICAL MASTER | VERTICAL MASTER | VERTICAL MASTER | VERTICAL MASTER | LINE MASTERS ⇩

HORIZONTAL MASTER

| # | | | | | | | # | | | | | | | # | | | | | | | # | | | | | | |
|---|---|---|---|---|---|---|---|---|---|---|---|---|---|---|---|---|---|---|---|---|---|---|---|---|---|---|---|
| 1 | 5 | 4 | 1 | 2 | 5 | 0 | 17 | 5 | 4 | 5 | 2 | 5 | 0 | 33 | 5 | 4 | 7 | 2 | 5 | 0 | 49 | 5 | 4 | 9 | 2 | 5 | 0 |
| 2 | 5 | 4 | 1 | 2 | 5 | 2 | 18 | 5 | 4 | 5 | 2 | 5 | 2 | 34 | 5 | 4 | 7 | 2 | 5 | 2 | 50 | 5 | 4 | 9 | 2 | 5 | 2 |
| 3 | 5 | 4 | 1 | 2 | 5 | 4 | 19 | 5 | 4 | 5 | 2 | 5 | 4 | 35 | 5 | 4 | 7 | 2 | 5 | 4 | 51 | 5 | 4 | 9 | 2 | 5 | 4 |
| 4 | 5 | 4 | 1 | 2 | 5 | 6 | 20 | 5 | 4 | 5 | 2 | 5 | 6 | 36 | 5 | 4 | 7 | 2 | 5 | 6 | 52 | 5 | 4 | 9 | 2 | 5 | 6 |

BLOCK MASTER | BLOCK MASTER | BLOCK MASTER | BLOCK MASTER

HORIZONTAL MASTER

| # | | | | | | | # | | | | | | | # | | | | | | | # | | | | | | |
|---|---|---|---|---|---|---|---|---|---|---|---|---|---|---|---|---|---|---|---|---|---|---|---|---|---|---|---|
| 5 | 5 | 4 | 1 | 4 | 5 | 0 | 21 | 5 | 4 | 5 | 4 | 5 | 0 | 37 | 5 | 4 | 7 | 4 | 5 | 0 | 53 | 5 | 4 | 9 | 4 | 5 | 0 |
| 6 | 5 | 4 | 1 | 4 | 5 | 2 | 22 | 5 | 4 | 5 | 4 | 5 | 2 | 38 | 5 | 4 | 7 | 4 | 5 | 2 | 54 | 5 | 4 | 9 | 4 | 5 | 2 |
| 7 | 5 | 4 | 1 | 4 | 5 | 4 | 23 | 5 | 4 | 5 | 4 | 5 | 4 | 39 | 5 | 4 | 7 | 4 | 5 | 4 | 55 | 5 | 4 | 9 | 4 | 5 | 4 |
| 8 | 5 | 4 | 1 | 4 | 5 | 6 | 24 | 5 | 4 | 5 | 4 | 5 | 6 | 40 | 5 | 4 | 7 | 4 | 5 | 6 | 56 | 5 | 4 | 9 | 4 | 5 | 6 |

BLOCK MASTER | BLOCK MASTER | BLOCK MASTER | BLOCK MASTER

HORIZONTAL MASTER

| # | | | | | | | # | | | | | | | # | | | | | | | # | | | | | | |
|---|---|---|---|---|---|---|---|---|---|---|---|---|---|---|---|---|---|---|---|---|---|---|---|---|---|---|---|
| 9 | 5 | 4 | 1 | 6 | 5 | 0 | 25 | 5 | 4 | 5 | 6 | 5 | 0 | 41 | 5 | 4 | 7 | 6 | 5 | 0 | 57 | 5 | 4 | 9 | 6 | 5 | 0 |
| 10 | 5 | 4 | 1 | 6 | 5 | 2 | 26 | 5 | 4 | 5 | 6 | 5 | 2 | 42 | 5 | 4 | 7 | 6 | 5 | 2 | 58 | 5 | 4 | 9 | 6 | 5 | 2 |
| 11 | 5 | 4 | 1 | 6 | 5 | 4 | 27 | 5 | 4 | 5 | 6 | 5 | 4 | 43 | 5 | 4 | 7 | 6 | 5 | 4 | 59 | 5 | 4 | 9 | 6 | 5 | 4 |
| 12 | 5 | 4 | 1 | 6 | 5 | 6 | 28 | 5 | 4 | 5 | 6 | 5 | 6 | 44 | 5 | 4 | 7 | 6 | 5 | 6 | 60 | 5 | 4 | 9 | 6 | 5 | 6 |

BLOCK MASTER | BLOCK MASTER | BLOCK MASTER | BLOCK MASTER

HORIZONTAL MASTER

| # | | | | | | | # | | | | | | | # | | | | | | | # | | | | | | |
|---|---|---|---|---|---|---|---|---|---|---|---|---|---|---|---|---|---|---|---|---|---|---|---|---|---|---|---|
| 13 | 5 | 4 | 1 | 8 | 5 | 0 | 29 | 5 | 4 | 5 | 8 | 5 | 0 | 45 | 5 | 4 | 7 | 8 | 5 | 0 | 61 | 5 | 4 | 9 | 8 | 5 | 0 |
| 14 | 5 | 4 | 1 | 8 | 5 | 2 | 30 | 5 | 4 | 5 | 8 | 5 | 2 | 46 | 5 | 4 | 7 | 8 | 5 | 2 | 62 | 5 | 4 | 9 | 8 | 5 | 2 |
| 15 | 5 | 4 | 1 | 8 | 5 | 4 | 31 | 5 | 4 | 5 | 8 | 5 | 4 | 47 | 5 | 4 | 7 | 8 | 5 | 4 | 63 | 5 | 4 | 9 | 8 | 5 | 4 |
| 16 | 5 | 4 | 1 | 8 | 5 | 6 | 32 | 5 | 4 | 5 | 8 | 5 | 6 | 48 | 5 | 4 | 7 | 8 | 5 | 6 | 64 | 5 | 4 | 9 | 8 | 5 | 6 |

BLOCK MASTER | BLOCK MASTER | BLOCK MASTER | BLOCK MASTER

| 5 | 4 | 3 | 0 | 5 | 8 | THEORETICAL MASTER |
|---|---|---|---|---|---|---|
| 3 | 2 | 1 | 2 | 7 | 0 | BASE FIELD (KBA) KEY BITTING ARRAY |
| 7 | 6 | 5 | 4 | 9 | 2 | |
| 9 | 8 | 7 | 6 | 1 | 4 | |
| 1 | 0 | 9 | 8 | 3 | 6 | |
| C | C | P | P | P | C | ⇦ SEQUENCE OF PROGRESSION |

VERTICAL MASTER | VERTICAL MASTER | VERTICAL MASTER | VERTICAL MASTER | LINE MASTERS ⇩

HORIZONTAL MASTER

| # | | | | | | | # | | | | | | | # | | | | | | | # | | | | | | |
|---|---|---|---|---|---|---|---|---|---|---|---|---|---|---|---|---|---|---|---|---|---|---|---|---|---|---|---|
| 1 | 5 | 4 | 1 | 2 | 7 | 8 | 17 | 5 | 4 | 5 | 2 | 7 | 8 | 33 | 5 | 4 | 7 | 2 | 7 | 8 | 49 | 5 | 4 | 9 | 2 | 7 | 8 |
| 2 | 5 | 4 | 1 | 2 | 9 | 8 | 18 | 5 | 4 | 5 | 2 | 9 | 8 | 34 | 5 | 4 | 7 | 2 | 9 | 8 | 50 | 5 | 4 | 9 | 2 | 9 | 8 |
| 3 | 5 | 4 | 1 | 2 | 1 | 8 | 19 | 5 | 4 | 5 | 2 | 1 | 8 | 35 | 5 | 4 | 7 | 2 | 1 | 8 | 51 | 5 | 4 | 9 | 2 | 1 | 8 |
| 4 | 5 | 4 | 1 | 2 | 3 | 8 | 20 | 5 | 4 | 5 | 2 | 3 | 8 | 36 | 5 | 4 | 7 | 2 | 3 | 8 | 52 | 5 | 4 | 9 | 2 | 3 | 8 |

BLOCK MASTER | BLOCK MASTER | BLOCK MASTER | BLOCK MASTER

HORIZONTAL MASTER

| # | | | | | | | # | | | | | | | # | | | | | | | # | | | | | | |
|---|---|---|---|---|---|---|---|---|---|---|---|---|---|---|---|---|---|---|---|---|---|---|---|---|---|---|---|
| 5 | 5 | 4 | 1 | 4 | 7 | 8 | 21 | 5 | 4 | 5 | 4 | 7 | 8 | 37 | 5 | 4 | 7 | 4 | 7 | 8 | 53 | 5 | 4 | 9 | 4 | 7 | 8 |
| 6 | 5 | 4 | 1 | 4 | 9 | 8 | 22 | 5 | 4 | 5 | 4 | 9 | 8 | 38 | 5 | 4 | 7 | 4 | 9 | 8 | 54 | 5 | 4 | 9 | 4 | 9 | 8 |
| 7 | 5 | 4 | 1 | 4 | 1 | 8 | 23 | 5 | 4 | 5 | 4 | 1 | 8 | 39 | 5 | 4 | 7 | 4 | 1 | 8 | 55 | 5 | 4 | 9 | 4 | 1 | 8 |
| 8 | 5 | 4 | 1 | 4 | 3 | 8 | 24 | 5 | 4 | 5 | 4 | 3 | 8 | 40 | 5 | 4 | 7 | 4 | 3 | 8 | 56 | 5 | 4 | 9 | 4 | 3 | 8 |

BLOCK MASTER | BLOCK MASTER | BLOCK MASTER | BLOCK MASTER

HORIZONTAL MASTER

| # | | | | | | | # | | | | | | | # | | | | | | | # | | | | | | |
|---|---|---|---|---|---|---|---|---|---|---|---|---|---|---|---|---|---|---|---|---|---|---|---|---|---|---|---|
| 9 | 5 | 4 | 1 | 6 | 7 | 8 | 25 | 5 | 4 | 5 | 6 | 7 | 8 | 41 | 5 | 4 | 7 | 6 | 7 | 8 | 57 | 5 | 4 | 9 | 6 | 7 | 8 |
| 10 | 5 | 4 | 1 | 6 | 9 | 8 | 26 | 5 | 4 | 5 | 6 | 9 | 8 | 42 | 5 | 4 | 7 | 6 | 9 | 8 | 58 | 5 | 4 | 9 | 6 | 9 | 8 |
| 11 | 5 | 4 | 1 | 6 | 1 | 8 | 27 | 5 | 4 | 5 | 6 | 1 | 8 | 43 | 5 | 4 | 7 | 6 | 1 | 8 | 59 | 5 | 4 | 9 | 6 | 1 | 8 |
| 12 | 5 | 4 | 1 | 6 | 3 | 8 | 28 | 5 | 4 | 5 | 6 | 3 | 8 | 44 | 5 | 4 | 7 | 6 | 3 | 8 | 60 | 5 | 4 | 9 | 6 | 3 | 8 |

BLOCK MASTER | BLOCK MASTER | BLOCK MASTER | BLOCK MASTER

HORIZONTAL MASTER

| # | | | | | | | # | | | | | | | # | | | | | | | # | | | | | | |
|---|---|---|---|---|---|---|---|---|---|---|---|---|---|---|---|---|---|---|---|---|---|---|---|---|---|---|---|
| 13 | 5 | 4 | 1 | 8 | 7 | 8 | 29 | 5 | 4 | 5 | 8 | 7 | 8 | 45 | 5 | 4 | 7 | 8 | 7 | 8 | 61 | 5 | 4 | 9 | 8 | 7 | 8 |
| 14 | 5 | 4 | 1 | 8 | 9 | 8 | 30 | 5 | 4 | 5 | 8 | 9 | 8 | 46 | 5 | 4 | 7 | 8 | 9 | 8 | 62 | 5 | 4 | 9 | 8 | 9 | 8 |
| 15 | 5 | 4 | 1 | 8 | 1 | 8 | 31 | 5 | 4 | 5 | 8 | 1 | 8 | 47 | 5 | 4 | 7 | 8 | 1 | 8 | 63 | 5 | 4 | 9 | 8 | 1 | 8 |
| 16 | 5 | 4 | 1 | 8 | 3 | 8 | 32 | 5 | 4 | 5 | 8 | 3 | 8 | 48 | 5 | 4 | 7 | 8 | 3 | 8 | 64 | 5 | 4 | 9 | 8 | 3 | 8 |

BLOCK MASTER | BLOCK MASTER | BLOCK MASTER | BLOCK MASTER

| | | | | | | |
|---|---|---|---|---|---|---|
| 5 | 4 | 3 | 0 | 5 | 8 | THEORETICAL MASTEF |
| 3 | 2 | 1 | 2 | 7 | 0 | BASE FIELD (KBA) KEY BITTING ARRAY |
| 7 | 6 | 5 | 4 | 9 | 2 | |
| 9 | 8 | 7 | 6 | 1 | 4 | |
| 1 | 0 | 9 | 8 | 3 | 6 | |
| C | P | C | C | P | P | ⇦ SEQUENCE OF PROGRESSION |

| | # | VERTICAL MASTER | # | VERTICAL MASTER | # | VERTICAL MASTER | # | VERTICAL MASTER | LINE MASTERS ⇩ |
|---|---|---|---|---|---|---|---|---|---|
| | 1 | 5 2 3 0 7 0 | 17 | 5 6 3 0 7 0 | 33 | 5 8 3 0 7 0 | 49 | 5 0 3 0 7 0 | |
| | 2 | 5 2 3 0 7 2 | 18 | 5 6 3 0 7 2 | 34 | 5 8 3 0 7 2 | 50 | 5 0 3 0 7 2 | |
| HORIZONTAL MASTER | 3 | 5 2 3 0 7 4 | 19 | 5 6 3 0 7 4 | 35 | 5 8 3 0 7 4 | 51 | 5 0 3 0 7 4 | |
| | 4 | 5 2 3 0 7 6 | 20 | 5 6 3 0 7 6 | 36 | 5 8 3 0 7 6 | 52 | 5 0 3 0 7 6 | |
| | | BLOCK MASTER | | BLOCK MASTER | | BLOCK MASTER | | BLOCK MASTER | |
| | 5 | 5 2 3 0 9 0 | 21 | 5 6 3 0 9 0 | 37 | 5 8 3 0 9 0 | 53 | 5 0 3 0 9 0 | |
| | 6 | 5 2 3 0 9 2 | 22 | 5 6 3 0 9 2 | 38 | 5 8 3 0 9 2 | 54 | 5 0 3 0 9 2 | |
| HORIZONTAL MASTER | 7 | 5 2 3 0 9 4 | 23 | 5 6 3 0 9 4 | 39 | 5 8 3 0 9 4 | 55 | 5 0 3 0 9 4 | |
| | 8 | 5 2 3 0 9 6 | 24 | 5 6 3 0 9 6 | 40 | 5 8 3 0 9 6 | 56 | 5 0 3 0 9 6 | |
| | | BLOCK MASTER | | BLOCK MASTER | | BLOCK MASTER | | BLOCK MASTER | |
| | 9 | 5 2 3 0 1 0 | 25 | 5 6 3 0 1 0 | 41 | 5 8 3 0 1 0 | 57 | 5 0 3 0 1 0 | |
| | 10 | 5 2 3 0 1 2 | 26 | 5 6 3 0 1 2 | 42 | 5 8 3 0 1 2 | 58 | 5 0 3 0 1 2 | |
| HORIZONTAL MASTER | 11 | 5 2 3 0 1 4 | 27 | 5 6 3 0 1 4 | 43 | 5 8 3 0 1 4 | 59 | 5 0 3 0 1 4 | |
| | 12 | 5 2 3 0 1 6 | 28 | 5 6 3 0 1 6 | 44 | 5 8 3 0 1 6 | 60 | 5 0 3 0 1 6 | |
| | | BLOCK MASTER | | BLOCK MASTER | | BLOCK MASTER | | BLOCK MASTER | |
| | 13 | 5 2 3 0 3 0 | 29 | 5 6 3 0 3 0 | 45 | 5 8 3 0 3 0 | 61 | 5 0 3 0 3 0 | |
| | 14 | 5 2 3 0 3 2 | 30 | 5 6 3 0 3 2 | 46 | 5 8 3 0 3 2 | 62 | 5 0 3 0 3 2 | |
| HORIZONTAL MASTER | 15 | 5 2 3 0 3 4 | 31 | 5 6 3 0 3 4 | 47 | 5 8 3 0 3 4 | 63 | 5 0 3 0 3 4 | |
| | 16 | 5 2 3 0 3 6 | 32 | 5 6 3 0 3 6 | 48 | 5 8 3 0 3 6 | 64 | 5 0 3 0 3 6 | |
| | | BLOCK MASTER | | BLOCK MASTER | | BLOCK MASTER | | BLOCK MASTER | |

| | | | | | | |
|---|---|---|---|---|---|---|
| 5 | 4 | 3 | 0 | 5 | 8 | THEORETICAL MASTER |
| 3 | 2 | 1 | 2 | 7 | 0 | BASE FIELD (KBA) KEY BITTING ARRAY |
| 7 | 6 | 5 | 4 | 9 | 2 | |
| 9 | 8 | 7 | 6 | 1 | 4 | |
| 1 | 0 | 9 | 8 | 3 | 6 | |
| C | P | C | P | C | P | ⇦ SEQUENCE OF PROGRESSION |

| | # | VERTICAL MASTER | # | VERTICAL MASTER | # | VERTICAL MASTER | # | VERTICAL MASTER | LINE MASTERS ⇩ |
|---|---|---|---|---|---|---|---|---|---|
| | 1 | 5 2 3 2 5 0 | 17 | 5 6 3 2 5 0 | 33 | 5 8 3 2 5 0 | 49 | 5 0 3 2 5 0 | |
| | 2 | 5 2 3 2 5 2 | 18 | 5 6 3 2 5 2 | 34 | 5 8 3 2 5 2 | 50 | 5 0 3 2 5 2 | |
| HORIZONTAL MASTER | 3 | 5 2 3 2 5 4 | 19 | 5 6 3 2 5 4 | 35 | 5 8 3 2 5 4 | 51 | 5 0 3 2 5 4 | |
| | 4 | 5 2 3 2 5 6 | 20 | 5 6 3 2 5 6 | 36 | 5 8 3 2 5 6 | 52 | 5 0 3 2 5 6 | |
| | | BLOCK MASTER | | BLOCK MASTER | | BLOCK MASTER | | BLOCK MASTER | |
| | 5 | 5 2 3 4 5 0 | 21 | 5 6 3 4 5 0 | 37 | 5 8 3 4 5 0 | 53 | 5 0 3 4 5 0 | |
| | 6 | 5 2 3 4 5 2 | 22 | 5 6 3 4 5 2 | 38 | 5 8 3 4 5 2 | 54 | 5 0 3 4 5 2 | |
| HORIZONTAL MASTER | 7 | 5 2 3 4 5 4 | 23 | 5 6 3 4 5 4 | 39 | 5 8 3 4 5 4 | 55 | 5 0 3 4 5 4 | |
| | 8 | 5 2 3 4 5 6 | 24 | 5 6 3 4 5 6 | 40 | 5 8 3 4 5 6 | 56 | 5 0 3 4 5 6 | |
| | | BLOCK MASTER | | BLOCK MASTER | | BLOCK MASTER | | BLOCK MASTER | |
| | 9 | 5 2 3 6 5 0 | 25 | 5 6 3 6 5 0 | 41 | 5 8 3 6 5 0 | 57 | 5 0 3 6 5 0 | |
| | 10 | 5 2 3 6 5 2 | 26 | 5 6 3 6 5 2 | 42 | 5 8 3 6 5 2 | 58 | 5 0 3 6 5 2 | |
| HORIZONTAL MASTER | 11 | 5 2 3 6 5 4 | 27 | 5 6 3 6 5 4 | 43 | 5 8 3 6 5 4 | 59 | 5 0 3 6 5 4 | |
| | 12 | 5 2 3 6 5 6 | 28 | 5 6 3 6 5 6 | 44 | 5 8 3 6 5 6 | 60 | 5 0 3 6 5 6 | |
| | | BLOCK MASTER | | BLOCK MASTER | | BLOCK MASTER | | BLOCK MASTER | |
| | 13 | 5 2 3 8 5 0 | 29 | 5 6 3 8 5 0 | 45 | 5 8 3 8 5 0 | 61 | 5 0 3 8 5 0 | |
| | 14 | 5 2 3 8 5 2 | 30 | 5 6 3 8 5 2 | 46 | 5 8 3 8 5 2 | 62 | 5 0 3 8 5 2 | |
| HORIZONTAL MASTER | 15 | 5 2 3 8 5 4 | 31 | 5 6 3 8 5 4 | 47 | 5 8 3 8 5 4 | 63 | 5 0 3 8 5 4 | |
| | 16 | 5 2 3 8 5 6 | 32 | 5 6 3 8 5 6 | 48 | 5 8 3 8 5 6 | 64 | 5 0 3 8 5 6 | |
| | | BLOCK MASTER | | BLOCK MASTER | | BLOCK MASTER | | BLOCK MASTER | |

| 5 | 4 | 3 | 0 | 5 | 8 | THEORETICAL MASTER |
|---|---|---|---|---|---|---|
| 3 | 2 | 1 | 2 | 7 | 0 | BASE FIELD (KBA) KEY BITTING ARRAY |
| 7 | 6 | 5 | 4 | 9 | 2 | |
| 9 | 8 | 7 | 6 | 1 | 4 | |
| 1 | 0 | 9 | 8 | 3 | 6 | |
| C | P | C | P | P | C | ⇦ SEQUENCE OF PROGRESSION |

| | | VERTICAL MASTER | | VERTICAL MASTER | | VERTICAL MASTER | | VERTICAL MASTER | LINE MASTERS ⇩ |
|---|---|---|---|---|---|---|---|---|---|
| | 1 | 5 2 3 2 7 8 | 17 | 5 6 3 2 7 8 | 33 | 5 8 3 2 7 8 | 49 | 5 0 3 2 7 8 | |
| | 2 | 5 2 3 2 9 8 | 18 | 5 6 3 2 9 8 | 34 | 5 8 3 2 9 8 | 50 | 5 0 3 2 9 8 | |
| HORIZONTAL MASTER | 3 | 5 2 3 2 1 8 | 19 | 5 6 3 2 1 8 | 35 | 5 8 3 2 1 8 | 51 | 5 0 3 2 1 8 | |
| | 4 | 5 2 3 2 3 8 | 20 | 5 6 3 2 3 8 | 36 | 5 8 3 2 3 8 | 52 | 5 0 3 2 3 8 | |
| | | BLOCK MASTER | | BLOCK MASTER | | BLOCK MASTER | | BLOCK MASTER | |
| | 5 | 5 2 3 4 7 8 | 21 | 5 6 3 4 7 8 | 37 | 5 8 3 4 7 8 | 53 | 5 0 3 4 7 8 | |
| | 6 | 5 2 3 4 9 8 | 22 | 5 6 3 4 9 8 | 38 | 5 8 3 4 9 8 | 54 | 5 0 3 4 9 8 | |
| HORIZONTAL MASTER | 7 | 5 2 3 4 1 8 | 23 | 5 6 3 4 1 8 | 39 | 5 8 3 4 1 8 | 55 | 5 0 3 4 1 8 | |
| | 8 | 5 2 3 4 3 8 | 24 | 5 6 3 4 3 8 | 40 | 5 8 3 4 3 8 | 56 | 5 0 3 4 3 8 | |
| | | BLOCK MASTER | | BLOCK MASTER | | BLOCK MASTER | | BLOCK MASTER | |
| | 9 | 5 2 3 6 7 8 | 25 | 5 6 3 6 7 8 | 41 | 5 8 3 6 7 8 | 57 | 5 0 3 6 7 8 | |
| | 10 | 5 2 3 6 9 8 | 26 | 5 6 3 6 9 8 | 42 | 5 8 3 6 9 8 | 58 | 5 0 3 6 9 8 | |
| HORIZONTAL MASTER | 11 | 5 2 3 6 1 8 | 27 | 5 6 3 6 1 8 | 43 | 5 8 3 6 1 8 | 59 | 5 0 3 6 1 8 | |
| | 12 | 5 2 3 6 3 8 | 28 | 5 6 3 6 3 8 | 44 | 5 8 3 6 3 8 | 60 | 5 0 3 6 3 8 | |
| | | BLOCK MASTER | | BLOCK MASTER | | BLOCK MASTER | | BLOCK MASTER | |
| | 13 | 5 2 3 8 7 8 | 29 | 5 6 3 8 7 8 | 45 | 5 8 3 8 7 8 | 61 | 5 0 3 8 7 8 | |
| | 14 | 5 2 3 8 9 8 | 30 | 5 6 3 8 9 8 | 46 | 5 8 3 8 9 8 | 62 | 5 0 3 8 9 8 | |
| HORIZONTAL MASTER | 15 | 5 2 3 8 1 8 | 31 | 5 6 3 8 1 8 | 47 | 5 8 3 8 1 8 | 63 | 5 0 3 8 1 8 | |
| | 16 | 5 2 3 8 3 8 | 32 | 5 6 3 8 3 8 | 48 | 5 8 3 8 3 8 | 64 | 5 0 3 8 3 8 | |
| | | BLOCK MASTER | | BLOCK MASTER | | BLOCK MASTER | | BLOCK MASTER | |

| 5 | 4 | 3 | 0 | 5 | 8 | THEORETICAL MASTER |
|---|---|---|---|---|---|---|
| 3 | 2 | 1 | 2 | 7 | 0 | BASE FIELD (KBA) KEY BITTING ARRAY |
| 7 | 6 | 5 | 4 | 9 | 2 | |
| 9 | 8 | 7 | 6 | 1 | 4 | |
| 1 | 0 | 9 | 8 | 3 | 6 | |
| C | P | P | C | C | P | ⇦ SEQUENCE OF PROGRESSION |

| | | VERTICAL MASTER | | VERTICAL MASTER | | VERTICAL MASTER | | VERTICAL MASTER | LINE MASTERS ⇩ |
|---|---|---|---|---|---|---|---|---|---|
| | 1 | 5 2 1 0 5 0 | 17 | 5 6 1 0 5 0 | 33 | 5 8 1 0 5 0 | 49 | 5 0 1 0 5 0 | |
| | 2 | 5 2 1 0 5 2 | 18 | 5 6 1 0 5 2 | 34 | 5 8 1 0 5 2 | 50 | 5 0 1 0 5 2 | |
| HORIZONTAL MASTER | 3 | 5 2 1 0 5 4 | 19 | 5 6 1 0 5 4 | 35 | 5 8 1 0 5 4 | 51 | 5 0 1 0 5 4 | |
| | 4 | 5 2 1 0 5 6 | 20 | 5 6 1 0 5 6 | 36 | 5 8 1 0 5 6 | 52 | 5 0 1 0 5 6 | |
| | | BLOCK MASTER | | BLOCK MASTER | | BLOCK MASTER | | BLOCK MASTER | |
| | 5 | 5 2 5 0 5 0 | 21 | 5 6 5 0 5 0 | 37 | 5 8 5 0 5 0 | 53 | 5 0 5 0 5 0 | |
| | 6 | 5 2 5 0 5 2 | 22 | 5 6 5 0 5 2 | 38 | 5 8 5 0 5 2 | 54 | 5 0 5 0 5 2 | |
| HORIZONTAL MASTER | 7 | 5 2 5 0 5 4 | 23 | 5 6 5 0 5 4 | 39 | 5 8 5 0 5 4 | 55 | 5 0 5 0 5 4 | |
| | 8 | 5 2 5 0 5 6 | 24 | 5 6 5 0 5 6 | 40 | 5 8 5 0 5 6 | 56 | 5 0 5 0 5 6 | |
| | | BLOCK MASTER | | BLOCK MASTER | | BLOCK MASTER | | BLOCK MASTER | |
| | 9 | 5 2 7 0 5 0 | 25 | 5 6 7 0 5 0 | 41 | 5 8 7 0 5 0 | 57 | 5 0 7 0 5 0 | |
| | 10 | 5 2 7 0 5 2 | 26 | 5 6 7 0 5 2 | 42 | 5 8 7 0 5 2 | 58 | 5 0 7 0 5 2 | |
| HORIZONTAL MASTER | 11 | 5 2 7 0 5 4 | 27 | 5 6 7 0 5 4 | 43 | 5 8 7 0 5 4 | 59 | 5 0 7 0 5 4 | |
| | 12 | 5 2 7 0 5 6 | 28 | 5 6 7 0 5 6 | 44 | 5 8 7 0 5 6 | 60 | 5 0 7 0 5 6 | |
| | | BLOCK MASTER | | BLOCK MASTER | | BLOCK MASTER | | BLOCK MASTER | |
| | 13 | 5 2 9 0 5 0 | 29 | 5 6 9 0 5 0 | 45 | 5 8 9 0 5 0 | 61 | 5 0 9 0 5 0 | |
| | 14 | 5 2 9 0 5 2 | 30 | 5 6 9 0 5 2 | 46 | 5 8 9 0 5 2 | 62 | 5 0 9 0 5 2 | |
| HORIZONTAL MASTER | 15 | 5 2 9 0 5 4 | 31 | 5 6 9 0 5 4 | 47 | 5 8 9 0 5 4 | 63 | 5 0 9 0 5 4 | |
| | 16 | 5 2 9 0 5 6 | 32 | 5 6 9 0 5 6 | 48 | 5 8 9 0 5 6 | 64 | 5 0 9 0 5 6 | |
| | | BLOCK MASTER | | BLOCK MASTER | | BLOCK MASTER | | BLOCK MASTER | |

| | | | | | | |
|---|---|---|---|---|---|---|
| 5 | 4 | 3 | 0 | 5 | 8 | THEORETICAL MASTER |
| 3 | 2 | 1 | 2 | 7 | 0 | BASE FIELD (KBA) KEY BITTING ARRAY |
| 7 | 6 | 5 | 4 | 9 | 2 | |
| 9 | 8 | 7 | 6 | 1 | 4 | |
| 1 | 0 | 9 | 8 | 3 | 6 | |
| C | P | P | C | P | C | ⇦ SEQUENCE OF PROGRESSION |

| | No. | VERTICAL MASTER | No. | VERTICAL MASTER | No. | VERTICAL MASTER | No. | VERTICAL MASTER | LINE MASTERS ⇩ |
|---|---|---|---|---|---|---|---|---|---|
| | 1 | 5 2 1 0 7 8 | 17 | 5 6 1 0 7 8 | 33 | 5 8 1 0 7 8 | 49 | 5 0 1 0 7 8 | |
| | 2 | 5 2 1 0 9 8 | 18 | 5 6 1 0 9 8 | 34 | 5 8 1 0 9 8 | 50 | 5 0 1 0 9 8 | |
| HORIZONTAL MASTER | 3 | 5 2 1 0 1 8 | 19 | 5 6 1 0 1 8 | 35 | 5 8 1 0 1 8 | 51 | 5 0 1 0 1 8 | |
| | 4 | 5 2 1 0 3 8 | 20 | 5 6 1 0 3 8 | 36 | 5 8 1 0 3 8 | 52 | 5 0 1 0 3 8 | |
| | | BLOCK MASTER | | BLOCK MASTER | | BLOCK MASTER | | BLOCK MASTER | |
| | 5 | 5 2 5 0 7 8 | 21 | 5 6 5 0 7 8 | 37 | 5 8 5 0 7 8 | 53 | 5 0 5 0 7 8 | |
| | 6 | 5 2 5 0 9 8 | 22 | 5 6 5 0 9 8 | 38 | 5 8 5 0 9 8 | 54 | 5 0 5 0 9 8 | |
| HORIZONTAL MASTER | 7 | 5 2 5 0 1 8 | 23 | 5 6 5 0 1 8 | 39 | 5 8 5 0 1 8 | 55 | 5 0 5 0 1 8 | |
| | 8 | 5 2 5 0 3 8 | 24 | 5 6 5 0 3 8 | 40 | 5 8 5 0 3 8 | 56 | 5 0 5 0 3 8 | |
| | | BLOCK MASTER | | BLOCK MASTER | | BLOCK MASTER | | BLOCK MASTER | |
| | 9 | 5 2 7 0 7 8 | 25 | 5 6 7 0 7 8 | 41 | 5 8 7 0 7 8 | 57 | 5 0 7 0 7 8 | |
| | 10 | 5 2 7 0 9 8 | 26 | 5 6 7 0 9 8 | 42 | 5 8 7 0 9 8 | 58 | 5 0 7 0 9 8 | |
| HORIZONTAL MASTER | 11 | 5 2 7 0 1 8 | 27 | 5 6 7 0 1 8 | 43 | 5 8 7 0 1 8 | 59 | 5 0 7 0 1 8 | |
| | 12 | 5 2 7 0 3 8 | 28 | 5 6 7 0 3 8 | 44 | 5 8 7 0 3 8 | 60 | 5 0 7 0 3 8 | |
| | | BLOCK MASTER | | BLOCK MASTER | | BLOCK MASTER | | BLOCK MASTER | |
| | 13 | 5 2 9 0 7 8 | 29 | 5 6 9 0 7 8 | 45 | 5 8 9 0 7 8 | 61 | 5 0 9 0 7 8 | |
| | 14 | 5 2 9 0 9 8 | 30 | 5 6 9 0 9 8 | 46 | 5 8 9 0 9 8 | 62 | 5 0 9 0 9 8 | |
| HORIZONTAL MASTER | 15 | 5 2 9 0 1 8 | 31 | 5 6 9 0 1 8 | 47 | 5 8 9 0 1 8 | 63 | 5 0 9 0 1 8 | |
| | 16 | 5 2 9 0 3 8 | 32 | 5 6 9 0 3 8 | 48 | 5 8 9 0 3 8 | 64 | 5 0 9 0 3 8 | |
| | | BLOCK MASTER | | BLOCK MASTER | | BLOCK MASTER | | BLOCK MASTER | |

| | | | | | | |
|---|---|---|---|---|---|---|
| 5 | 4 | 3 | 0 | 5 | 8 | THEORETICAL MASTER |
| 3 | 2 | 1 | 2 | 7 | 0 | BASE FIELD (KBA) KEY BITTING ARRAY |
| 7 | 6 | 5 | 4 | 9 | 2 | |
| 9 | 8 | 7 | 6 | 1 | 4 | |
| 1 | 0 | 9 | 8 | 3 | 6 | |
| C | P | P | P | C | C | ⇦ SEQUENCE OF PROGRESSION |

| | No. | VERTICAL MASTER | No. | VERTICAL MASTER | No. | VERTICAL MASTER | No. | VERTICAL MASTER | LINE MASTERS ⇩ |
|---|---|---|---|---|---|---|---|---|---|
| | 1 | 5 2 1 2 5 8 | 17 | 5 6 1 2 5 8 | 33 | 5 8 1 2 5 8 | 49 | 5 0 1 2 5 8 | |
| | 2 | 5 2 1 4 5 8 | 18 | 5 6 1 4 5 8 | 34 | 5 8 1 4 5 8 | 50 | 5 0 1 4 5 8 | |
| HORIZONTAL MASTER | 3 | 5 2 1 6 5 8 | 19 | 5 6 1 6 5 8 | 35 | 5 8 1 6 5 8 | 51 | 5 0 1 6 5 8 | |
| | 4 | 5 2 1 8 5 8 | 20 | 5 6 1 8 5 8 | 36 | 5 8 1 8 5 8 | 52 | 5 0 1 8 5 8 | |
| | | BLOCK MASTER | | BLOCK MASTER | | BLOCK MASTER | | BLOCK MASTER | |
| | 5 | 5 2 5 2 5 8 | 21 | 5 6 5 2 5 8 | 37 | 5 8 5 2 5 8 | 53 | 5 0 5 2 5 8 | |
| | 6 | 5 2 5 4 5 8 | 22 | 5 6 5 4 5 8 | 38 | 5 8 5 4 5 8 | 54 | 5 0 5 4 5 8 | |
| HORIZONTAL MASTER | 7 | 5 2 5 6 5 8 | 23 | 5 6 5 6 5 8 | 39 | 5 8 5 6 5 8 | 55 | 5 0 5 6 5 8 | |
| | 8 | 5 2 5 8 5 8 | 24 | 5 6 5 8 5 8 | 40 | 5 8 5 8 5 8 | 56 | 5 0 5 8 5 8 | |
| | | BLOCK MASTER | | BLOCK MASTER | | BLOCK MASTER | | BLOCK MASTER | |
| | 9 | 5 2 7 2 5 8 | 25 | 5 6 7 2 5 8 | 41 | 5 8 7 2 5 8 | 57 | 5 0 7 2 5 8 | |
| | 10 | 5 2 7 4 5 8 | 26 | 5 6 7 4 5 8 | 42 | 5 8 7 4 5 8 | 58 | 5 0 7 4 5 8 | |
| HORIZONTAL MASTER | 11 | 5 2 7 6 5 8 | 27 | 5 6 7 6 5 8 | 43 | 5 8 7 6 5 8 | 59 | 5 0 7 6 5 8 | |
| | 12 | 5 2 7 8 5 8 | 28 | 5 6 7 8 5 8 | 44 | 5 8 7 8 5 8 | 60 | 5 0 7 8 5 8 | |
| | | BLOCK MASTER | | BLOCK MASTER | | BLOCK MASTER | | BLOCK MASTER | |
| | 13 | 5 2 9 2 5 8 | 29 | 5 6 9 2 5 8 | 45 | 5 8 9 2 5 8 | 61 | 5 0 9 2 5 8 | |
| | 14 | 5 2 9 4 5 8 | 30 | 5 6 9 4 5 8 | 46 | 5 8 9 4 5 8 | 62 | 5 0 9 4 5 8 | |
| HORIZONTAL MASTER | 15 | 5 2 9 6 5 8 | 31 | 5 6 9 6 5 8 | 47 | 5 8 9 6 5 8 | 63 | 5 0 9 6 5 8 | |
| | 16 | 5 2 9 8 5 8 | 32 | 5 6 9 8 5 8 | 48 | 5 8 9 8 5 8 | 64 | 5 0 9 8 5 8 | |
| | | BLOCK MASTER | | BLOCK MASTER | | BLOCK MASTER | | BLOCK MASTER | |

| | | | | | | |
|---|---|---|---|---|---|---|
| 5 | 4 | 3 | 0 | 5 | 8 | THEORETICAL MASTER |
| 3 | 2 | 1 | 2 | 7 | 0 | BASE FIELD (KBA) KEY BITTING ARRAY |
| 7 | 6 | 5 | 4 | 9 | 2 | |
| 9 | 8 | 7 | 6 | 1 | 4 | |
| 1 | 0 | 9 | 8 | 3 | 6 | |
| P | C | C | C | P | P | ⇦ SEQUENCE OF PROGRESSION |

| | # | VERTICAL MASTER | # | VERTICAL MASTER | # | VERTICAL MASTER | # | VERTICAL MASTER | LINE MASTERS ⇩ |
|---|---|---|---|---|---|---|---|---|---|
| | 1 | 3 4 3 0 7 0 | 17 | 7 4 3 0 7 0 | 33 | 9 4 3 0 7 0 | 49 | 1 4 3 0 7 0 | |
| | 2 | 3 4 3 0 7 2 | 18 | 7 4 3 0 7 2 | 34 | 9 4 3 0 7 2 | 50 | 1 4 3 0 7 2 | |
| HORIZONTAL MASTER | 3 | 3 4 3 0 7 4 | 19 | 7 4 3 0 7 4 | 35 | 9 4 3 0 7 4 | 51 | 1 4 3 0 7 4 | |
| | 4 | 3 4 3 0 7 6 | 20 | 7 4 3 0 7 6 | 36 | 9 4 3 0 7 6 | 52 | 1 4 3 0 7 6 | |

| | # | BLOCK MASTER | # | BLOCK MASTER | # | BLOCK MASTER | # | BLOCK MASTER | |
|---|---|---|---|---|---|---|---|---|---|
| | 5 | 3 4 3 0 9 0 | 21 | 7 4 3 0 9 0 | 37 | 9 4 3 0 9 0 | 53 | 1 4 3 0 9 0 | |
| | 6 | 3 4 3 0 9 2 | 22 | 7 4 3 0 9 2 | 38 | 9 4 3 0 9 2 | 54 | 1 4 3 0 9 2 | |
| HORIZONTAL MASTER | 7 | 3 4 3 0 9 4 | 23 | 7 4 3 0 9 4 | 39 | 9 4 3 0 9 4 | 55 | 1 4 3 0 9 4 | |
| | 8 | 3 4 3 0 9 6 | 24 | 7 4 3 0 9 6 | 40 | 9 4 3 0 9 6 | 56 | 1 4 3 0 9 6 | |

| | # | BLOCK MASTER | # | BLOCK MASTER | # | BLOCK MASTER | # | BLOCK MASTER | |
|---|---|---|---|---|---|---|---|---|---|
| | 9 | 3 4 3 0 1 0 | 25 | 7 4 3 0 1 0 | 41 | 9 4 3 0 1 0 | 57 | 1 4 3 0 1 0 | |
| | 10 | 3 4 3 0 1 2 | 26 | 7 4 3 0 1 2 | 42 | 9 4 3 0 1 2 | 58 | 1 4 3 0 1 2 | |
| HORIZONTAL MASTER | 11 | 3 4 3 0 1 4 | 27 | 7 4 3 0 1 4 | 43 | 9 4 3 0 1 4 | 59 | 1 4 3 0 1 4 | |
| | 12 | 3 4 3 0 1 6 | 28 | 7 4 3 0 1 6 | 44 | 9 4 3 0 1 6 | 60 | 1 4 3 0 1 6 | |

| | # | BLOCK MASTER | # | BLOCK MASTER | # | BLOCK MASTER | # | BLOCK MASTER | |
|---|---|---|---|---|---|---|---|---|---|
| | 13 | 3 4 3 0 3 0 | 29 | 7 4 3 0 3 0 | 45 | 9 4 3 0 3 0 | 61 | 1 4 3 0 3 0 | |
| | 14 | 3 4 3 0 3 2 | 30 | 7 4 3 0 3 2 | 46 | 9 4 3 0 3 2 | 62 | 1 4 3 0 3 2 | |
| HORIZONTAL MASTER | 15 | 3 4 3 0 3 4 | 31 | 7 4 3 0 3 4 | 47 | 9 4 3 0 3 4 | 63 | 1 4 3 0 3 4 | |
| | 16 | 3 4 3 0 3 6 | 32 | 7 4 3 0 3 6 | 48 | 9 4 3 0 3 6 | 64 | 1 4 3 0 3 6 | |

BLOCK MASTER BLOCK MASTER BLOCK MASTER BLOCK MASTER

| | | | | | | |
|---|---|---|---|---|---|---|
| 5 | 4 | 3 | 0 | 5 | 8 | THEORETICAL MASTER |
| 3 | 2 | 1 | 2 | 7 | 0 | BASE FIELD (KBA) KEY BITTING ARRAY |
| 7 | 6 | 5 | 4 | 9 | 2 | |
| 9 | 8 | 7 | 6 | 1 | 4 | |
| 1 | 0 | 9 | 8 | 3 | 6 | |
| P | C | C | P | C | P | ⇦ SEQUENCE OF PROGRESSION |

| | # | VERTICAL MASTER | # | VERTICAL MASTER | # | VERTICAL MASTER | # | VERTICAL MASTER | LINE MASTERS ⇩ |
|---|---|---|---|---|---|---|---|---|---|
| | 1 | 3 4 3 2 5 0 | 17 | 7 4 3 2 5 0 | 33 | 9 4 3 2 5 0 | 49 | 1 4 3 2 5 0 | |
| | 2 | 3 4 3 2 5 2 | 18 | 7 4 3 2 5 2 | 34 | 9 4 3 2 5 2 | 50 | 1 4 3 2 5 2 | |
| HORIZONTAL MASTER | 3 | 3 4 3 2 5 4 | 19 | 7 4 3 2 5 4 | 35 | 9 4 3 2 5 4 | 51 | 1 4 3 2 5 4 | |
| | 4 | 3 4 3 2 5 6 | 20 | 7 4 3 2 5 6 | 36 | 9 4 3 2 5 6 | 52 | 1 4 3 2 5 6 | |

| | # | BLOCK MASTER | # | BLOCK MASTER | # | BLOCK MASTER | # | BLOCK MASTER | |
|---|---|---|---|---|---|---|---|---|---|
| | 5 | 3 4 3 4 5 0 | 21 | 7 4 3 4 5 0 | 37 | 9 4 3 4 5 0 | 53 | 1 4 3 4 5 0 | |
| | 6 | 3 4 3 4 5 2 | 22 | 7 4 3 4 5 2 | 38 | 9 4 3 4 5 2 | 54 | 1 4 3 4 5 2 | |
| HORIZONTAL MASTER | 7 | 3 4 3 4 5 4 | 23 | 7 4 3 4 5 4 | 39 | 9 4 3 4 5 4 | 55 | 1 4 3 4 5 4 | |
| | 8 | 3 4 3 4 5 6 | 24 | 7 4 3 4 5 6 | 40 | 9 4 3 4 5 6 | 56 | 1 4 3 4 5 6 | |

| | # | BLOCK MASTER | # | BLOCK MASTER | # | BLOCK MASTER | # | BLOCK MASTER | |
|---|---|---|---|---|---|---|---|---|---|
| | 9 | 3 4 3 6 5 0 | 25 | 7 4 3 6 5 0 | 41 | 9 4 3 6 5 0 | 57 | 1 4 3 6 5 0 | |
| | 10 | 3 4 3 6 5 2 | 26 | 7 4 3 6 5 2 | 42 | 9 4 3 6 5 2 | 58 | 1 4 3 6 5 2 | |
| HORIZONTAL MASTER | 11 | 3 4 3 6 5 4 | 27 | 7 4 3 6 5 4 | 43 | 9 4 3 6 5 4 | 59 | 1 4 3 6 5 4 | |
| | 12 | 3 4 3 6 5 6 | 28 | 7 4 3 6 5 6 | 44 | 9 4 3 6 5 6 | 60 | 1 4 3 6 5 6 | |

| | # | BLOCK MASTER | # | BLOCK MASTER | # | BLOCK MASTER | # | BLOCK MASTER | |
|---|---|---|---|---|---|---|---|---|---|
| | 13 | 3 4 3 8 5 0 | 29 | 7 4 3 8 5 0 | 45 | 9 4 3 8 5 0 | 61 | 1 4 3 8 5 0 | |
| | 14 | 3 4 3 8 5 2 | 30 | 7 4 3 8 5 2 | 46 | 9 4 3 8 5 2 | 62 | 1 4 3 8 5 2 | |
| HORIZONTAL MASTER | 15 | 3 4 3 8 5 4 | 31 | 7 4 3 8 5 4 | 47 | 9 4 3 8 5 4 | 63 | 1 4 3 8 5 4 | |
| | 16 | 3 4 3 8 5 6 | 32 | 7 4 3 8 5 6 | 48 | 9 4 3 8 5 6 | 64 | 1 4 3 8 5 6 | |

BLOCK MASTER BLOCK MASTER BLOCK MASTER BLOCK MASTER

| 5 | 4 | 3 | 0 | 5 | 8 | THEORETICAL MASTER |
|---|---|---|---|---|---|---|
| 3 | 2 | 1 | 2 | 7 | 0 | BASE FIELD (KBA) KEY BITTING ARRAY |
| 7 | 6 | 5 | 4 | 9 | 2 | |
| 9 | 8 | 7 | 6 | 1 | 4 | |
| 1 | 0 | 9 | 8 | 3 | 6 | |
| P | C | C | P | P | C | ⇦ SEQUENCE OF PROGRESSION |

VERTICAL MASTER — VERTICAL MASTER — VERTICAL MASTER — VERTICAL MASTER — LINE MASTERS ⇩

HORIZONTAL MASTER

| | | | | | | | | LINE MASTERS |
|---|---|---|---|---|---|---|---|---|
| 1 | 3 4 3 2 7 8 | 17 | 7 4 3 2 7 8 | 33 | 9 4 3 2 7 8 | 49 | 1 4 3 2 7 8 | |
| 2 | 3 4 3 2 9 8 | 18 | 7 4 3 2 9 8 | 34 | 9 4 3 2 9 8 | 50 | 1 4 3 2 9 8 | |
| 3 | 3 4 3 2 1 8 | 19 | 7 4 3 2 1 8 | 35 | 9 4 3 2 1 8 | 51 | 1 4 3 2 1 8 | |
| 4 | 3 4 3 2 3 8 | 20 | 7 4 3 2 3 8 | 36 | 9 4 3 2 3 8 | 52 | 1 4 3 2 3 8 | |

BLOCK MASTER — BLOCK MASTER — BLOCK MASTER — BLOCK MASTER

HORIZONTAL MASTER

| | | | | | | | | |
|---|---|---|---|---|---|---|---|---|
| 5 | 3 4 3 4 7 8 | 21 | 7 4 3 4 7 8 | 37 | 9 4 3 4 7 8 | 53 | 1 4 3 4 7 8 | |
| 6 | 3 4 3 4 9 8 | 22 | 7 4 3 4 9 8 | 38 | 9 4 3 4 9 8 | 54 | 1 4 3 4 9 8 | |
| 7 | 3 4 3 4 1 8 | 23 | 7 4 3 4 1 8 | 39 | 9 4 3 4 1 8 | 55 | 1 4 3 4 1 8 | |
| 8 | 3 4 3 4 3 8 | 24 | 7 4 3 4 3 8 | 40 | 9 4 3 4 3 8 | 56 | 1 4 3 4 3 8 | |

BLOCK MASTER — BLOCK MASTER — BLOCK MASTER — BLOCK MASTER

HORIZONTAL MASTER

| | | | | | | | | |
|---|---|---|---|---|---|---|---|---|
| 9 | 3 4 3 6 7 8 | 25 | 7 4 3 6 7 8 | 41 | 9 4 3 6 7 8 | 57 | 1 4 3 6 7 8 | |
| 10 | 3 4 3 6 9 8 | 26 | 7 4 3 6 9 8 | 42 | 9 4 3 6 9 8 | 58 | 1 4 3 6 9 8 | |
| 11 | 3 4 3 6 1 8 | 27 | 7 4 3 6 1 8 | 43 | 9 4 3 6 1 8 | 59 | 1 4 3 6 1 8 | |
| 12 | 3 4 3 6 3 8 | 28 | 7 4 3 6 3 8 | 44 | 9 4 3 6 3 8 | 60 | 1 4 3 6 3 8 | |

BLOCK MASTER — BLOCK MASTER — BLOCK MASTER — BLOCK MASTER

HORIZONTAL MASTER

| | | | | | | | | |
|---|---|---|---|---|---|---|---|---|
| 13 | 3 4 3 8 7 8 | 29 | 7 4 3 8 7 8 | 45 | 9 4 3 8 7 8 | 61 | 1 4 3 8 7 8 | |
| 14 | 3 4 3 8 9 8 | 30 | 7 4 3 8 9 8 | 46 | 9 4 3 8 9 8 | 62 | 1 4 3 8 9 8 | |
| 15 | 3 4 3 8 1 8 | 31 | 7 4 3 8 1 8 | 47 | 9 4 3 8 1 8 | 63 | 1 4 3 8 1 8 | |
| 16 | 3 4 3 8 3 8 | 32 | 7 4 3 8 3 8 | 48 | 9 4 3 8 3 8 | 64 | 1 4 3 8 3 8 | |

| 5 | 4 | 3 | 0 | 5 | 8 | THEORETICAL MASTER |
|---|---|---|---|---|---|---|
| 3 | 2 | 1 | 2 | 7 | 0 | BASE FIELD (KBA) KEY BITTING ARRAY |
| 7 | 6 | 5 | 4 | 9 | 2 | |
| 9 | 8 | 7 | 6 | 1 | 4 | |
| 1 | 0 | 9 | 8 | 3 | 6 | |
| P | C | P | C | C | P | ⇦ SEQUENCE OF PROGRESSION |

VERTICAL MASTER — VERTICAL MASTER — VERTICAL MASTER — VERTICAL MASTER — LINE MASTERS ⇩

HORIZONTAL MASTER

| | | | | | | | | LINE MASTERS |
|---|---|---|---|---|---|---|---|---|
| 1 | 3 4 1 0 5 0 | 17 | 7 4 1 0 5 0 | 33 | 9 4 1 0 5 0 | 49 | 1 4 1 0 5 0 | |
| 2 | 3 4 1 0 5 2 | 18 | 7 4 1 0 5 2 | 34 | 9 4 1 0 5 2 | 50 | 1 4 1 0 5 2 | |
| 3 | 3 4 1 0 5 4 | 19 | 7 4 1 0 5 4 | 35 | 9 4 1 0 5 4 | 51 | 1 4 1 0 5 4 | |
| 4 | 3 4 1 0 5 6 | 20 | 7 4 1 0 5 6 | 36 | 9 4 1 0 5 6 | 52 | 1 4 1 0 5 6 | |

BLOCK MASTER — BLOCK MASTER — BLOCK MASTER — BLOCK MASTER

HORIZONTAL MASTER

| | | | | | | | | |
|---|---|---|---|---|---|---|---|---|
| 5 | 3 4 5 0 5 0 | 21 | 7 4 5 0 5 0 | 37 | 9 4 5 0 5 0 | 53 | 1 4 5 0 5 0 | |
| 6 | 3 4 5 0 5 2 | 22 | 7 4 5 0 5 2 | 38 | 9 4 5 0 5 2 | 54 | 1 4 5 0 5 2 | |
| 7 | 3 4 5 0 5 4 | 23 | 7 4 5 0 5 4 | 39 | 9 4 5 0 5 4 | 55 | 1 4 5 0 5 4 | |
| 8 | 3 4 5 0 5 6 | 24 | 7 4 5 0 5 6 | 40 | 9 4 5 0 5 6 | 56 | 1 4 5 0 5 6 | |

BLOCK MASTER — BLOCK MASTER — BLOCK MASTER — BLOCK MASTER

HORIZONTAL MASTER

| | | | | | | | | |
|---|---|---|---|---|---|---|---|---|
| 9 | 3 4 7 0 5 0 | 25 | 7 4 7 0 5 0 | 41 | 9 4 7 0 5 0 | 57 | 1 4 7 0 5 0 | |
| 10 | 3 4 7 0 5 2 | 26 | 7 4 7 0 5 2 | 42 | 9 4 7 0 5 2 | 58 | 1 4 7 0 5 2 | |
| 11 | 3 4 7 0 5 4 | 27 | 7 4 7 0 5 4 | 43 | 9 4 7 0 5 4 | 59 | 1 4 7 0 5 4 | |
| 12 | 3 4 7 0 5 6 | 28 | 7 4 7 0 5 6 | 44 | 9 4 7 0 5 6 | 60 | 1 4 7 0 5 6 | |

BLOCK MASTER — BLOCK MASTER — BLOCK MASTER — BLOCK MASTER

HORIZONTAL MASTER

| | | | | | | | | |
|---|---|---|---|---|---|---|---|---|
| 13 | 3 4 9 0 5 0 | 29 | 7 4 9 0 5 0 | 45 | 9 4 9 0 5 0 | 61 | 1 4 9 0 5 0 | |
| 14 | 3 4 9 0 5 2 | 30 | 7 4 9 0 5 2 | 46 | 9 4 9 0 5 2 | 62 | 1 4 9 0 5 2 | |
| 15 | 3 4 9 0 5 4 | 31 | 7 4 9 0 5 4 | 47 | 9 4 9 0 5 4 | 63 | 1 4 9 0 5 4 | |
| 16 | 3 4 9 0 5 6 | 32 | 7 4 9 0 5 6 | 48 | 9 4 9 0 5 6 | 64 | 1 4 9 0 5 6 | |

BLOCK MASTER — BLOCK MASTER — BLOCK MASTER — BLOCK MASTER

| | | | | | | |
|---|---|---|---|---|---|---|
| 5 | 4 | 3 | 0 | 5 | 8 | THEORETICAL MASTER |
| 3 | 2 | 1 | 2 | 7 | 0 | BASE FIELD (KBA) KEY BITTING ARRAY |
| 7 | 6 | 5 | 4 | 9 | 2 | |
| 9 | 8 | 7 | 6 | 1 | 4 | |
| 1 | 0 | 9 | 8 | 3 | 6 | |
| P | C | P | C | P | C | ⇦ SEQUENCE OF PROGRESSION |

| | # | | # | | # | | # | | LINE MASTERS ⇩ |
|---|---|---|---|---|---|---|---|---|---|
| | | VERTICAL MASTER | | VERTICAL MASTER | | VERTICAL MASTER | | VERTICAL MASTER | |
| | 1 | 3 4 1 0 7 8 | 17 | 7 4 1 0 7 8 | 33 | 9 4 1 0 7 8 | 49 | 1 4 1 0 7 8 | |
| | 2 | 3 4 1 0 9 8 | 18 | 7 4 1 0 9 8 | 34 | 9 4 1 0 9 8 | 50 | 1 4 1 0 9 8 | |
| HORIZONTAL MASTER | 3 | 3 4 1 0 1 8 | 19 | 7 4 1 0 1 8 | 35 | 9 4 1 0 1 8 | 51 | 1 4 1 0 1 8 | |
| | 4 | 3 4 1 0 3 8 | 20 | 7 4 1 0 3 8 | 36 | 9 4 1 0 3 8 | 52 | 1 4 1 0 3 8 | |
| | | BLOCK MASTER | | BLOCK MASTER | | BLOCK MASTER | | BLOCK MASTER | |
| | 5 | 3 4 5 0 7 8 | 21 | 7 4 5 0 7 8 | 37 | 9 4 5 0 7 8 | 53 | 1 4 5 0 7 8 | |
| | 6 | 3 4 5 0 9 8 | 22 | 7 4 5 0 9 8 | 38 | 9 4 5 0 9 8 | 54 | 1 4 5 0 9 8 | |
| HORIZONTAL MASTER | 7 | 3 4 5 0 1 8 | 23 | 7 4 5 0 1 8 | 39 | 9 4 5 0 1 8 | 55 | 1 4 5 0 1 8 | |
| | 8 | 3 4 5 0 3 8 | 24 | 7 4 5 0 3 8 | 40 | 9 4 5 0 3 8 | 56 | 1 4 5 0 3 8 | |
| | | BLOCK MASTER | | BLOCK MASTER | | BLOCK MASTER | | BLOCK MASTER | |
| | 9 | 3 4 7 0 7 8 | 25 | 7 4 7 0 7 8 | 41 | 9 4 7 0 7 8 | 57 | 1 4 7 0 7 8 | |
| | 10 | 3 4 7 0 9 8 | 26 | 7 4 7 0 9 8 | 42 | 9 4 7 0 9 8 | 58 | 1 4 7 0 9 8 | |
| HORIZONTAL MASTER | 11 | 3 4 7 0 1 8 | 27 | 7 4 7 0 1 8 | 43 | 9 4 7 0 1 8 | 59 | 1 4 7 0 1 8 | |
| | 12 | 3 4 7 0 3 8 | 28 | 7 4 7 0 3 8 | 44 | 9 4 7 0 3 8 | 60 | 1 4 7 0 3 8 | |
| | | BLOCK MASTER | | BLOCK MASTER | | BLOCK MASTER | | BLOCK MASTER | |
| | 13 | 3 4 9 0 7 8 | 29 | 7 4 9 0 7 8 | 45 | 9 4 9 0 7 8 | 61 | 1 4 9 0 7 8 | |
| | 14 | 3 4 9 0 9 8 | 30 | 7 4 9 0 9 8 | 46 | 9 4 9 0 9 8 | 62 | 1 4 9 0 9 8 | |
| HORIZONTAL MASTER | 15 | 3 4 9 0 1 8 | 31 | 7 4 9 0 1 8 | 47 | 9 4 9 0 1 8 | 63 | 1 4 9 0 1 8 | |
| | 16 | 3 4 9 0 3 8 | 32 | 7 4 9 0 3 8 | 48 | 9 4 9 0 3 8 | 64 | 1 4 9 0 3 8 | |
| | | BLOCK MASTER | | BLOCK MASTER | | BLOCK MASTER | | BLOCK MASTER | |

| | | | | | | |
|---|---|---|---|---|---|---|
| 5 | 4 | 3 | 0 | 5 | 8 | THEORETICAL MASTER |
| 3 | 2 | 1 | 2 | 7 | 0 | BASE FIELD (KBA) KEY BITTING ARRAY |
| 7 | 6 | 5 | 4 | 9 | 2 | |
| 9 | 8 | 7 | 6 | 1 | 4 | |
| 1 | 0 | 9 | 8 | 3 | 6 | |
| P | C | P | P | C | C | ⇦ SEQUENCE OF PROGRESSION |

| | # | | # | | # | | # | | LINE MASTERS ⇩ |
|---|---|---|---|---|---|---|---|---|---|
| | | VERTICAL MASTER | | VERTICAL MASTER | | VERTICAL MASTER | | VERTICAL MASTER | |
| | 1 | 3 4 1 2 5 8 | 17 | 7 4 1 2 5 8 | 33 | 9 4 1 2 5 8 | 49 | 1 4 1 2 5 8 | |
| | 2 | 3 4 1 4 5 8 | 18 | 7 4 1 4 5 8 | 34 | 9 4 1 4 5 8 | 50 | 1 4 1 4 5 8 | |
| HORIZONTAL MASTER | 3 | 3 4 1 6 5 8 | 19 | 7 4 1 6 5 8 | 35 | 9 4 1 6 5 8 | 51 | 1 4 1 6 5 8 | |
| | 4 | 3 4 1 8 5 8 | 20 | 7 4 1 8 5 8 | 36 | 9 4 1 8 5 8 | 52 | 1 4 1 8 5 8 | |
| | | BLOCK MASTER | | BLOCK MASTER | | BLOCK MASTER | | BLOCK MASTER | |
| | 5 | 3 4 5 2 5 8 | 21 | 7 4 5 2 5 8 | 37 | 9 4 5 2 5 8 | 53 | 1 4 5 2 5 8 | |
| | 6 | 3 4 5 4 5 8 | 22 | 7 4 5 4 5 8 | 38 | 9 4 5 4 5 8 | 54 | 1 4 5 4 5 8 | |
| HORIZONTAL MASTER | 7 | 3 4 5 6 5 8 | 23 | 7 4 5 6 5 8 | 39 | 9 4 5 6 5 8 | 55 | 1 4 5 6 5 8 | |
| | 8 | 3 4 5 8 5 8 | 24 | 7 4 5 8 5 8 | 40 | 9 4 5 8 5 8 | 56 | 1 4 5 8 5 8 | |
| | | BLOCK MASTER | | BLOCK MASTER | | BLOCK MASTER | | BLOCK MASTER | |
| | 9 | 3 4 7 2 5 8 | 25 | 7 4 7 2 5 8 | 41 | 9 4 7 2 5 8 | 57 | 1 4 7 2 5 8 | |
| | 10 | 3 4 7 4 5 8 | 26 | 7 4 7 4 5 8 | 42 | 9 4 7 4 5 8 | 58 | 1 4 7 4 5 8 | |
| HORIZONTAL MASTER | 11 | 3 4 7 6 5 8 | 27 | 7 4 7 6 5 8 | 43 | 9 4 7 6 5 8 | 59 | 1 4 7 6 5 8 | |
| | 12 | 3 4 7 8 5 8 | 28 | 7 4 7 8 5 8 | 44 | 9 4 7 8 5 8 | 60 | 1 4 7 8 5 8 | |
| | | BLOCK MASTER | | BLOCK MASTER | | BLOCK MASTER | | BLOCK MASTER | |
| | 13 | 3 4 9 2 5 8 | 29 | 7 4 9 2 5 8 | 45 | 9 4 9 2 5 8 | 61 | 1 4 9 2 5 8 | |
| | 14 | 3 4 9 4 5 8 | 30 | 7 4 9 4 5 8 | 46 | 9 4 9 4 5 8 | 62 | 1 4 9 4 5 8 | |
| HORIZONTAL MASTER | 15 | 3 4 9 6 5 8 | 31 | 7 4 9 6 5 8 | 47 | 9 4 9 6 5 8 | 63 | 1 4 9 6 5 8 | |
| | 16 | 3 4 9 8 5 8 | 32 | 7 4 9 8 5 8 | 48 | 9 4 9 8 5 8 | 64 | 1 4 9 8 5 8 | |
| | | BLOCK MASTER | | BLOCK MASTER | | BLOCK MASTER | | BLOCK MASTER | |

| | | | | | | |
|---|---|---|---|---|---|---|
| 5 | 4 | 3 | 0 | 5 | 8 | THEORETICAL MASTER |
| 3 | 2 | 1 | 2 | 7 | 0 | BASE FIELD (KBA) KEY BITTING ARRAY |
| 7 | 6 | 5 | 4 | 9 | 2 | |
| 9 | 8 | 7 | 6 | 1 | 4 | |
| 1 | 0 | 9 | 8 | 3 | 6 | |
| P | P | C | C | C | P | ⇦ SEQUENCE OF PROGRESSION |

| | No. | VERTICAL MASTER | No. | VERTICAL MASTER | No. | VERTICAL MASTER | No. | VERTICAL MASTER | LINE MASTERS ⇩ |
|---|---|---|---|---|---|---|---|---|---|
| | 1 | 323050 | 17 | 723050 | 33 | 923050 | 49 | 123050 | |
| | 2 | 323052 | 18 | 723052 | 34 | 923052 | 50 | 123052 | |
| HORIZONTAL MASTER | 3 | 323054 | 19 | 723054 | 35 | 923054 | 51 | 123054 | |
| | 4 | 323056 | 20 | 723056 | 36 | 923056 | 52 | 123056 | |
| | | BLOCK MASTER | | BLOCK MASTER | | BLOCK MASTER | | BLOCK MASTER | |
| | 5 | 363050 | 21 | 763050 | 37 | 963050 | 53 | 163050 | |
| | 6 | 363052 | 22 | 763052 | 38 | 963052 | 54 | 163052 | |
| HORIZONTAL MASTER | 7 | 363054 | 23 | 763054 | 39 | 963054 | 55 | 163054 | |
| | 8 | 363056 | 24 | 763056 | 40 | 963056 | 56 | 163056 | |
| | | BLOCK MASTER | | BLOCK MASTER | | BLOCK MASTER | | BLOCK MASTER | |
| | 9 | 383050 | 25 | 783050 | 41 | 983050 | 57 | 183050 | |
| | 10 | 383052 | 26 | 783052 | 42 | 983052 | 58 | 183052 | |
| HORIZONTAL MASTER | 11 | 383054 | 27 | 783054 | 43 | 983054 | 59 | 183054 | |
| | 12 | 383056 | 28 | 783056 | 44 | 983056 | 60 | 183056 | |
| | | BLOCK MASTER | | BLOCK MASTER | | BLOCK MASTER | | BLOCK MASTER | |
| | 13 | 303050 | 29 | 703050 | 45 | 903050 | 61 | 103050 | |
| | 14 | 303052 | 30 | 703052 | 46 | 903052 | 62 | 103052 | |
| HORIZONTAL MASTER | 15 | 303054 | 31 | 703054 | 47 | 903054 | 63 | 103054 | |
| | 16 | 303056 | 32 | 703056 | 48 | 903056 | 64 | 103056 | |
| | | BLOCK MASTER | | BLOCK MASTER | | BLOCK MASTER | | BLOCK MASTER | |

| | | | | | | |
|---|---|---|---|---|---|---|
| 5 | 4 | 3 | 0 | 5 | 8 | THEORETICAL MASTER |
| 3 | 2 | 1 | 2 | 7 | 0 | BASE FIELD (KBA) KEY BITTING ARRAY |
| 7 | 6 | 5 | 4 | 9 | 2 | |
| 9 | 8 | 7 | 6 | 1 | 4 | |
| 1 | 0 | 9 | 8 | 3 | 6 | |
| P | P | C | C | P | C | ⇦ SEQUENCE OF PROGRESSION |

| | No. | VERTICAL MASTER | No. | VERTICAL MASTER | No. | VERTICAL MASTER | No. | VERTICAL MASTER | LINE MASTERS ⇩ |
|---|---|---|---|---|---|---|---|---|---|
| | 1 | 323078 | 17 | 723078 | 33 | 923078 | 49 | 123078 | |
| | 2 | 323098 | 18 | 723098 | 34 | 923098 | 50 | 123098 | |
| HORIZONTAL MASTER | 3 | 323018 | 19 | 723018 | 35 | 923018 | 51 | 123018 | |
| | 4 | 323038 | 20 | 723038 | 36 | 923038 | 52 | 123038 | |
| | | BLOCK MASTER | | BLOCK MASTER | | BLOCK MASTER | | BLOCK MASTER | |
| | 5 | 363078 | 21 | 763078 | 37 | 963078 | 53 | 163078 | |
| | 6 | 363098 | 22 | 763098 | 38 | 963098 | 54 | 163098 | |
| HORIZONTAL MASTER | 7 | 363018 | 23 | 763018 | 39 | 963018 | 55 | 163018 | |
| | 8 | 363038 | 24 | 763038 | 40 | 963038 | 56 | 163038 | |
| | | BLOCK MASTER | | BLOCK MASTER | | BLOCK MASTER | | BLOCK MASTER | |
| | 9 | 383078 | 25 | 783078 | 41 | 983078 | 57 | 183078 | |
| | 10 | 383098 | 26 | 783098 | 42 | 983098 | 58 | 183098 | |
| HORIZONTAL MASTER | 11 | 383018 | 27 | 783018 | 43 | 983018 | 59 | 183018 | |
| | 12 | 383038 | 28 | 783038 | 44 | 983038 | 60 | 183038 | |
| | | BLOCK MASTER | | BLOCK MASTER | | BLOCK MASTER | | BLOCK MASTER | |
| | 13 | 303078 | 29 | 703078 | 45 | 903078 | 61 | 103078 | |
| | 14 | 303098 | 30 | 703098 | 46 | 903098 | 62 | 103098 | |
| HORIZONTAL MASTER | 15 | 303018 | 31 | 703018 | 47 | 903018 | 63 | 103018 | |
| | 16 | 303038 | 32 | 703038 | 48 | 903038 | 64 | 103038 | |
| | | BLOCK MASTER | | BLOCK MASTER | | BLOCK MASTER | | BLOCK MASTER | |

| | | | | | |
|---|---|---|---|---|---|
| 5 | 4 | 3 | 0 | 5 | 8 |
| 3 | 2 | 1 | 2 | 7 | 0 |
| 7 | 6 | 5 | 4 | 9 | 2 |
| 9 | 8 | 7 | 6 | 1 | 4 |
| 1 | 0 | 9 | 8 | 3 | 6 |
| P | P | C | P | C | C |

THEORETICAL MASTER (first row)

BASE FIELD (KBA) KEY BITTING ARRAY (rows 2–5)

SEQUENCE OF PROGRESSION (last row)

| | No. | VERTICAL MASTER | No. | VERTICAL MASTER | No. | VERTICAL MASTER | No. | VERTICAL MASTER | LINE MASTERS |
|---|---|---|---|---|---|---|---|---|---|
| | 1 | 3 2 3 2 5 8 | 17 | 7 2 3 2 5 8 | 33 | 9 2 3 2 5 8 | 49 | 1 2 3 2 5 8 | |
| | 2 | 3 2 3 4 5 8 | 18 | 7 2 3 4 5 8 | 34 | 9 2 3 4 5 8 | 50 | 1 2 3 4 5 8 | |
| HORIZONTAL MASTER | 3 | 3 2 3 6 5 8 | 19 | 7 2 3 6 5 8 | 35 | 9 2 3 6 5 8 | 51 | 1 2 3 6 5 8 | |
| | 4 | 3 2 3 8 5 8 | 20 | 7 2 3 8 5 8 | 36 | 9 2 3 8 5 8 | 52 | 1 2 3 8 5 8 | |
| | | BLOCK MASTER | | BLOCK MASTER | | BLOCK MASTER | | BLOCK MASTER | |
| | 5 | 3 6 3 2 5 8 | 21 | 7 6 3 2 5 8 | 37 | 9 6 3 2 5 8 | 53 | 1 6 3 2 5 8 | |
| | 6 | 3 6 3 4 5 8 | 22 | 7 6 3 4 5 8 | 38 | 9 6 3 4 5 8 | 54 | 1 6 3 4 5 8 | |
| HORIZONTAL MASTER | 7 | 3 6 3 6 5 8 | 23 | 7 6 3 6 5 8 | 39 | 9 6 3 6 5 8 | 55 | 1 6 3 6 5 8 | |
| | 8 | 3 6 3 8 5 8 | 24 | 7 6 3 8 5 8 | 40 | 9 6 3 8 5 8 | 56 | 1 6 3 8 5 8 | |
| | | BLOCK MASTER | | BLOCK MASTER | | BLOCK MASTER | | BLOCK MASTER | |
| | 9 | 3 8 3 2 5 8 | 25 | 7 8 3 2 5 8 | 41 | 9 8 3 2 5 8 | 57 | 1 8 3 2 5 8 | |
| | 10 | 3 8 3 4 5 8 | 26 | 7 8 3 4 5 8 | 42 | 9 8 3 4 5 8 | 58 | 1 8 3 4 5 8 | |
| HORIZONTAL MASTER | 11 | 3 8 3 6 5 8 | 27 | 7 8 3 6 5 8 | 43 | 9 8 3 6 5 8 | 59 | 1 8 3 6 5 8 | |
| | 12 | 3 8 3 8 5 8 | 28 | 7 8 3 8 5 8 | 44 | 9 8 3 8 5 8 | 60 | 1 8 3 8 5 8 | |
| | | BLOCK MASTER | | BLOCK MASTER | | BLOCK MASTER | | BLOCK MASTER | |
| | 13 | 3 0 3 2 5 8 | 29 | 7 0 3 2 5 8 | 45 | 9 0 3 2 5 8 | 61 | 1 0 3 2 5 8 | |
| | 14 | 3 0 3 4 5 8 | 30 | 7 0 3 4 5 8 | 46 | 9 0 3 4 5 8 | 62 | 1 0 3 4 5 8 | |
| HORIZONTAL MASTER | 15 | 3 0 3 6 5 8 | 31 | 7 0 3 6 5 8 | 47 | 9 0 3 6 5 8 | 63 | 1 0 3 6 5 8 | |
| | 16 | 3 0 3 8 5 8 | 32 | 7 0 3 8 5 8 | 48 | 9 0 3 8 5 8 | 64 | 1 0 3 8 5 8 | |
| | | BLOCK MASTER | | BLOCK MASTER | | BLOCK MASTER | | BLOCK MASTER | |

| | | | | | |
|---|---|---|---|---|---|
| 5 | 4 | 3 | 0 | 5 | 8 |
| 3 | 2 | 1 | 2 | 7 | 0 |
| 7 | 6 | 5 | 4 | 9 | 2 |
| 9 | 8 | 7 | 6 | 1 | 4 |
| 1 | 0 | 9 | 8 | 3 | 6 |
| P | P | P | C | C | C |

THEORETICAL MASTER (first row)

BASE FIELD (KBA) KEY BITTING ARRAY (rows 2–5)

SEQUENCE OF PROGRESSION (last row)

| | No. | VERTICAL MASTER | No. | VERTICAL MASTER | No. | VERTICAL MASTER | No. | VERTICAL MASTER | LINE MASTERS |
|---|---|---|---|---|---|---|---|---|---|
| | 1 | 3 2 1 0 5 8 | 17 | 7 2 1 0 5 8 | 33 | 9 2 1 0 5 8 | 49 | 1 2 1 0 5 8 | |
| | 2 | 3 2 5 0 5 8 | 18 | 7 2 5 0 5 8 | 34 | 9 2 5 0 5 8 | 50 | 1 2 5 0 5 8 | |
| HORIZONTAL MASTER | 3 | 3 2 7 0 5 8 | 19 | 7 2 7 0 5 8 | 35 | 9 2 7 0 5 8 | 51 | 1 2 7 0 5 8 | |
| | 4 | 3 2 9 0 5 8 | 20 | 7 2 9 0 5 8 | 36 | 9 2 9 0 5 8 | 52 | 1 2 9 0 5 8 | |
| | | BLOCK MASTER | | BLOCK MASTER | | BLOCK MASTER | | BLOCK MASTER | |
| | 5 | 3 6 1 0 5 8 | 21 | 7 6 1 0 5 8 | 37 | 9 6 1 0 5 8 | 53 | 1 6 1 0 5 8 | |
| | 6 | 3 6 5 0 5 8 | 22 | 7 6 5 0 5 8 | 38 | 9 6 5 0 5 8 | 54 | 1 6 5 0 5 8 | |
| HORIZONTAL MASTER | 7 | 3 6 7 0 5 8 | 23 | 7 6 7 0 5 8 | 39 | 9 6 7 0 5 8 | 55 | 1 6 7 0 5 8 | |
| | 8 | 3 6 9 0 5 8 | 24 | 7 6 9 0 5 8 | 40 | 9 6 9 0 5 8 | 56 | 1 6 9 0 5 8 | |
| | | BLOCK MASTER | | BLOCK MASTER | | BLOCK MASTER | | BLOCK MASTER | |
| | 9 | 3 8 1 0 5 8 | 25 | 7 8 1 0 5 8 | 41 | 9 8 1 0 5 8 | 57 | 1 8 1 0 5 8 | |
| | 10 | 3 8 5 0 5 8 | 26 | 7 8 5 0 5 8 | 42 | 9 8 5 0 5 8 | 58 | 1 8 5 0 5 8 | |
| HORIZONTAL MASTER | 11 | 3 8 7 0 5 8 | 27 | 7 8 7 0 5 8 | 43 | 9 8 7 0 5 8 | 59 | 1 8 7 0 5 8 | |
| | 12 | 3 8 9 0 5 8 | 28 | 7 8 9 0 5 8 | 44 | 9 8 9 0 5 8 | 60 | 1 8 9 0 5 8 | |
| | | BLOCK MASTER | | BLOCK MASTER | | BLOCK MASTER | | BLOCK MASTER | |
| | 13 | 3 0 1 0 5 8 | 29 | 7 0 1 0 5 8 | 45 | 9 0 1 0 5 8 | 61 | 1 0 1 0 5 8 | |
| | 14 | 3 0 5 0 5 8 | 30 | 7 0 5 0 5 8 | 46 | 9 0 5 0 5 8 | 62 | 1 0 5 0 5 8 | |
| HORIZONTAL MASTER | 15 | 3 0 7 0 5 8 | 31 | 7 0 7 0 5 8 | 47 | 9 0 7 0 5 8 | 63 | 1 0 7 0 5 8 | |
| | 16 | 3 0 9 0 5 8 | 32 | 7 0 9 0 5 8 | 48 | 9 0 9 0 5 8 | 64 | 1 0 9 0 5 8 | |
| | | BLOCK MASTER | | BLOCK MASTER | | BLOCK MASTER | | BLOCK MASTER | |

# Remember:

If there is **one** position being progressed, it will need a **Block** (4 or 5) to fully develop and will yield that 4 (or 5 in single increment) change key bittings per pattern.

If there are **two** positions being progressed, it will need a **Vertical Column** (16 or 25) to fully develop and will yield that 16 (or 25 in single increment) change key bittings per pattern.

If there are **three** positions being progressed, it will need a **Page of bittings** (64 or 125) to fully develop and will yield that 64 (or 125 in single increment) change key bittings per pattern.

If there are **four** positions being progressed, it will need a **Group of pages** (4 pages or 5 pages) to fully develop and will yield 256 (or 625 in single increment) change key bittings per pattern.

If there are **five** positions being progressed (on a 6 pin or more cylinder only, of course) , it will need a **section of pages** (16 pages or 25 pages) to fully develop and will yield 1,024 (or 3,125 in single increment) change key bittings per pattern.

If there are **six** positions being progressed (on a 7 pin or more cylinder only, of course), it will need a **6 pin system** (64 pages or 125 pages) to fully develop and will yield 4,096 (or 15,625 in single increment) change key bittings per pattern.

# CHAPTER SIX:

# ADDING MASTER KEYS FOR ROTATING CONSTANT SYSTEMS

The first thing to understand in breaking a Rotating Constant system into smaller master key groups is that the number of positions we are progressing will be the first limitation we face.

As mentioned in the previous lesson, when Progressing 1, the progression can be accomplished for each pattern on a Block. Therefore, the Theoretical Master is in reality a Block Master, and there are no submasters of any level except that Block Master.

When Progressing 2, the progression can be accomplished using a Vertical Column for each progression pattern, and there are 15 patterns for a six pin key. Therefore, the Theoretical Master is in reality a Vertical Master. The only submasters are the Block Masters. There are additional incidental master keys that operate the cylinder, but they are not clearly defined masters such as the block master. Each pattern yields 16 Change Keys, (25 for a single increment with 5 possibilities in the KBA) as it normally would under the Vertical Master.

When Progressing 3, the progression can be accomplished using a full page for each progression pattern, and there are 20 patterns for a six pin key. Therefore, the Theoretical Master is in reality a Page Master, and we have the capability of Block Masters, Vertical Masters, Horizontal Masters, and Line Masters under it, as well as a few undefined incidental master keys. Each Pattern yields 64 Change Keys (125 for a single increment with 5 possibilities in the KBA)

When Progressing 4, we will need a Group of pages to accomplish it. Each Pattern yields 256 Change Keys (625 for a single increment with 5 possibilities in the KBA)

Progressing 5 requires a Section of pages to accomplish each progression pattern. Each Pattern yields 1,024 Change Keys (3,125 for a single increment with 5 possibilities in the KBA)

Each Master must be filled in on the chart before any selection of appropriate masters is made, since it will then be discovered that what was a Block Master in one progression pattern of Progressing 3, for example, will also be a Line Master in another progression, another Block Master in another, and especially difficult to spot, may operate as a sort of Vertical Line Master in another.

This is because a Block Master in Holding 4 Progressing 2, for example, will end up having 5 out of six cuts in common with the Theoretical Master. It will be limited only by one cut position, which will give it its assigned level of authority within the system.
If it is a Block Master for the first block in the first column, everything else it is will also be a "first", e.g. the first line or the first combination in the each block in the first column.

In a progression pattern Progressing 3, the Block and Line Masters will follow that same Pattern, the Vertical and Horizontal Masters will typically still operate one column or none per chart, but may also have a different assignment – one combination in each of the sixteen blocks on a page. Once again, if they are the first column in one chart, for a chart on which they operate one bitting per block it will be the first combination. If they are the second column (horizontal or vertical) on one chart and there is another chart on which they operate one bitting per block, it will be the second bitting. They NEVER jump positions.

Let's look at the standard incidental masters for our Holding 3 constant of 6 pattern developed in the previous chapter.

The first progression pattern (CCCPPP) looked like this:

| | | | | | | |
|---|---|---|---|---|---|---|
| 5 | 4 | 3 | 0 | 5 | 8 | THEORETICAL MASTER |
| 3 | 2 | 1 | 2 | 7 | 0 | |
| 7 | 6 | 5 | 4 | 9 | 2 | BASE FIELD (KBA) KEY BITTING ARRAY |
| 9 | 8 | 7 | 6 | 1 | 4 | |
| 1 | 0 | 9 | 8 | 3 | 6 | |
| C | C | C | P | P | P | ⇦ SEQUENCE OF PROGRESSION |

PAGE MASTER ______
GROUP MASTER ______
SECTION MASTER ______

PAGE # ______ OF ______ PAGES

| | # | VERTICAL MASTER | # | VERTICAL MASTER | # | VERTICAL MASTER | # | VERTICAL MASTER | LINE MASTERS ⇩ |
|---|---|---|---|---|---|---|---|---|---|
| | 1 | 5 4 3 2 7 0 | 17 | 5 4 3 4 7 0 | 33 | 5 4 3 6 7 0 | 49 | 5 4 3 8 7 0 | |
| | 2 | 5 4 3 2 7 2 | 18 | 5 4 3 4 7 2 | 34 | 5 4 3 6 7 2 | 50 | 5 4 3 8 7 2 | |
| HORIZONTAL MASTER | 3 | 5 4 3 2 7 4 | 19 | 5 4 3 4 7 4 | 35 | 5 4 3 6 7 4 | 51 | 5 4 3 8 7 4 | |
| | 4 | 5 4 3 2 7 6 | 20 | 5 4 3 4 7 6 | 36 | 5 4 3 6 7 6 | 52 | 5 4 3 8 7 6 | |
| | | BLOCK MASTER | | BLOCK MASTER | | BLOCK MASTER | | BLOCK MASTER | |
| | 5 | 5 4 3 2 9 0 | 21 | 5 4 3 4 9 0 | 37 | 5 4 3 6 9 0 | 53 | 5 4 3 8 9 0 | |
| | 6 | 5 4 3 2 9 2 | 22 | 5 4 3 4 9 2 | 38 | 5 4 3 6 9 2 | 54 | 5 4 3 8 9 2 | |
| HORIZONTAL MASTER | 7 | 5 4 3 2 9 4 | 23 | 5 4 3 4 9 4 | 39 | 5 4 3 6 9 4 | 55 | 5 4 3 8 9 4 | |
| | 8 | 5 4 3 2 9 6 | 24 | 5 4 3 4 9 6 | 40 | 5 4 3 6 9 6 | 56 | 5 4 3 8 9 6 | |
| | | BLOCK MASTER | | BLOCK MASTER | | BLOCK MASTER | | BLOCK MASTER | |
| | 9 | 5 4 3 2 1 0 | 25 | 5 4 3 4 1 0 | 41 | 5 4 3 6 1 0 | 57 | 5 4 3 8 1 0 | |
| | 10 | 5 4 3 2 1 2 | 26 | 5 4 3 4 1 2 | 42 | 5 4 3 6 1 2 | 58 | 5 4 3 8 1 2 | |
| HORIZONTAL MASTER | 11 | 5 4 3 2 1 4 | 27 | 5 4 3 4 1 4 | 43 | 5 4 3 6 1 4 | 59 | 5 4 3 8 1 4 | |
| | 12 | 5 4 3 2 1 6 | 28 | 5 4 3 4 1 6 | 44 | 5 4 3 6 1 6 | 60 | 5 4 3 8 1 6 | |
| | | BLOCK MASTER | | BLOCK MASTER | | BLOCK MASTER | | BLOCK MASTER | |
| | 13 | 5 4 3 2 3 0 | 29 | 5 4 3 4 3 0 | 45 | 5 4 3 6 3 0 | 61 | 5 4 3 8 3 0 | |
| | 14 | 5 4 3 2 3 2 | 30 | 5 4 3 4 3 2 | 46 | 5 4 3 6 3 2 | 62 | 5 4 3 8 3 2 | |
| HORIZONTAL MASTER | 15 | 5 4 3 2 3 4 | 31 | 5 4 3 4 3 4 | 47 | 5 4 3 6 3 4 | 63 | 5 4 3 8 3 4 | |
| | 16 | 5 4 3 2 3 6 | 32 | 5 4 3 4 3 6 | 48 | 5 4 3 6 3 6 | 64 | 5 4 3 8 3 6 | |

Therefore, the first Block looks like this:

| | | | | | | |
|---|---|---|---|---|---|---|
| 1 | 5 | 4 | 3 | 2 | 7 | 0 |
| 2 | 5 | 4 | 3 | 2 | 7 | 2 |
| 3 | 5 | 4 | 3 | 2 | 7 | 4 |
| 4 | 5 | 4 | 3 | 2 | 7 | 6 |

BLOCK MASTER

The Block Master Key (BMK) would be 5 4 3 2 7 8 and each Block on that page will simply use what is common within the Block and use the cut from the TMK in the first progression sequence (no matter where the first progression sequence might be).

The Block Master Keys (BMK) for the first progression pattern (CCCPPP) are:

| | | | |
|---|---|---|---|
| 543278 | 543478 | 543678 | 543878 |
| 543298 | 543498 | 543698 | 543898 |
| 543218 | 543418 | 543618 | 543818 |
| 543238 | 543438 | 543638 | 543838 |

The first block on the second progression pattern (CCPCPP) looks like this, and once again the first progression sequence (first to change) is on the far right (5 4 1 0 7 8 ):

| | | | | | | |
|---|---|---|---|---|---|---|
| 1 | 5 | 4 | 1 | 0 | 7 | 0 |
| 2 | 5 | 4 | 1 | 0 | 7 | 2 |
| 3 | 5 | 4 | 1 | 0 | 7 | 4 |
| 4 | 5 | 4 | 1 | 0 | 7 | 6 |

BLOCK MASTER

The Block Master Keys on the second progression pattern of our example (CCPCPP) are:

| | | | |
|---|---|---|---|
| 541078 | 545078 | 547078 | 549078 |
| 541098 | 545098 | 547098 | 549098 |
| 541018 | 545018 | 547018 | 549018 |
| 541038 | 545038 | 547038 | 549038 |

The first block on the third progression pattern (CCPPCP) also has the first progression sequence on the far right. Why? Because if you examine all the above, they all have a P in the far right position, indicating the first progression sequence is in that position.

The Block Master Keys on the third progression pattern of our example (CCPPCP) are:

| 541258 | 545258 | 547258 | 549258 |
|---|---|---|---|
| 541458 | 545458 | 547458 | 549458 |
| 541658 | 545658 | 547658 | 549658 |
| 541858 | 545858 | 547858 | 549858 |

But if we jump ahead to the fourth progression sequence we showed above (CCPPPC), now we have a C in the far right position so the position next to it with the first P from the right is the first progression sequence:

| | | | | | | |
|---|---|---|---|---|---|---|
| 1 | 5 | 4 | 1 | 2 | 7 | 8 |
| 2 | 5 | 4 | 1 | 2 | 9 | 8 |
| 3 | 5 | 4 | 1 | 2 | 1 | 8 |
| 4 | 5 | 4 | 1 | 2 | 3 | 8 |

:

BLOCK MASTER

So in this case the BMK is 5 4 1 2 5 8. Remember that wherever the cut changes from bitting to bitting, that position is the first progression sequence and the Block Master will in that position use a TMK cut.

The Block Master Keys on the fourth progression pattern of our example (CCPPPC) are:

| 541258 | 545258 | 547258 | 549258 |
|---|---|---|---|
| 541458 | 545458 | 547458 | 549458 |
| 541658 | 545658 | 547658 | 549658 |
| 541858 | 545858 | 547858 | 549858 |

By this time you should have an idea how the Block Master Key selection is done, but let's just do one more pattern to make certain of it. The last progression pattern we showed above was PPPCCC. The third position from the left was the first to have changes occur, so it is our first progression sequence, and uses the TMK cut from that position for the Block Master Key.

The Block Master Keys on the last progression pattern of our example (PPPCCC) are:

| 323058 | 723058 | 923058 | 123058 |
|---|---|---|---|
| 363058 | 763058 | 963058 | 163058 |
| 383058 | 783058 | 983058 | 183858 |
| 303058 | 703058 | 903058 | 103058 |

Got it? Good. Now look again at the Block Master Keys for the third (CCPPCP) and the fourth (CCPPPC) progression patterns. Notice anything? The Block Master Key set for both these patterns is identical!

That is because the progression pattern from the four positions on the left is identical, and in the remaining positions, one uses the TMK cut in the second position from the right as a constant position and the other uses it as the BMK cut. In the position on the far right, the situation is reversed and the key that used the previous position as a constant now uses the BMK cut there, and the key that used the BMK cut there now uses it as a constant. Therefore, the Block Master Keys are the same for both progression patterns.

Now let's look at the Line Master Keys for progression pattern PCPPCC. They are:

541258
541458
541658
541858
545258
545458
545658
545858
547258
547458
547658
547858
549258
549458
549658
549858

Do they look familiar? Yes, they are the same as the Block Master Keys from patterns CCPPCP and CCPPPC. That is because the PCPPCC used a CC on the right two positions, creating the same position there as we saw on the first two positions, and the third progression sequence in that pattern (remember it is the third to have changes occur regardless of its position on the key and the Line Master Key uses the TMK cut from the third progression sequence.

Remember that we only have three progression sequences to work with, the first, second and third. In a TPP system, they would have corresponded to progression sequences a, b and c. There is no d, e or f progression equivalent, because our patterns are based on only progressing 3 positions out of the 6 on the key.

Our Block Master Key uses the TMK cut from position "a" (first progression sequence). Our Line Master Key uses the TMK cut from position "c" (third progression sequence.

The patterns we saw so far were CCPPCP, CCPPPC, (where it was a Block Master Key) and PCPPCC (where it was a Line Master Key). In pattern CPPPCC, we see that the two on the far right have the CC that gives us the 58 used by the master keys we saw so far, and the first P on the left is the third progression sequence, so once again the Line Master Keys will be the same as the set we have seen so far.

This is the reason many books, courses and software regarding master keying state that the incidental master keys cannot be used in an RCM system, except when using certain specific RCM techniques which allow you to add levels by "splitting the key".

But if you think about it logically, it is not all that difficult. Looking at these master keys, we can see that all of them have progressed positions in the two in the middle of the key. That limits them to four progression patterns: CPPPCC and PCPPCC, where it will be a Line Master Key, and CCPPPC and CCPPCP where it will be a Block Master Key.

Let's use this technique to predict where a particular master key will have authority. Let's use 543278. We see that it has two different from the TMK that will define it. This is because it is a Block Master Key somewhere and we are already holding three positions constant. Three constants plus one for the master key = 4 positions of 6, leaving 2 positions different.

Possibilities for the Progression Patterns where this master key has power are:
CCCPPP, CCPPPC, CPCPPC, PCCPPC. On the CCCPPP, it will be a Block Master Key because the position will be to the right of the other two positions, and therefore the first progression sequence. On the CCPPPC, the CPCPPC, and the PCCPPC, it will act as a Line Master because it will be in the third progression sequence. That really is not all that difficult to determine, is it?

Let's try another. Let's use 343078. This one will be a little trickier, because the two "different" cuts are not next to each other. When the two cuts were next to each other, they could ONLY be "ab" or "bc", which means the master involved could only use the a or c position. But in this case the P on the right could be "a" or "b" and the P on the left could only be a "c", which means the master key involved could use the "a" or "b" position. We can have the "a" as a Block Master on the far right, with pattern PCCCPP. We can have the "b" position on patterns PCCPPC, PCPCPC and PPCCPC. In each of these patterns it will have authority over the top bitting in each block in the first Vertical column.

Regardless of the positions involved, using one additional cut from the TMK will let it operate 4 bittings in a "Base Four" system, and there will be four patterns where it has authority...no more and no less. However, the number of patterns is because we are using two change key cuts to define the master key. If we had a typical single increment system, adding one more TMK cut to the three we are holding constant would let a key operate five bittings, but, using two change key positions out of six, there would still be four patterns where it appeared.

Ok, that covers the Block Master Key (which uses the TMK cut from the First Progression Sequence), the Line Master Key (which uses the TMK cut from the Third Progression Sequence) and what might be called a vertical line master key, but ever has been called that officially (which uses the TMK cut from the Second Progression Sequence.

Let's go up a level. We already should know how to assign the bitting for the Vertical Column Master Key and the Horizontal Master Key, and know that on a two increment drop ("Base Four" KBA) each master of this sort will operate 16 different change key bittings (25 on a "Base Five" KBA such as a typical single increment drop).

The Vertical Master Key (VMK) uses TMK cuts "a" and "b" (the first and second progression sequences) and the Horizontal Master Key uses cuts "a" and "c" (first and third progression sequences). There is also the capability for a master key which uses cuts "b" and "c" (second and third progression sequences) and which will have authority over one bitting per block in every block on a progression sequence page.

The first VMK from progression pattern CCCPPP is 543258. Its authority will be limited to bittings that have a 2 in the third position from the right. Therefore, it must have a P in that position for it to have any authority.

On CCCPPP it is a Vertical Master Key because the 2 is in the third progression sequence("c") and the master is using progression sequences "a" and "b" (first and second).

On CCPPCP the 2 is in the second progression sequence ("b") so the master is using progression sequences "a" and "c" (first and third) so it will be a Horizontal Master Key.

On CCPPPC, CPCPCP and CPCPPC, the 2 is once again in the second position of the progressed positions, making the master a Horizontal Master Key ("a" and "c") using the first and third positions.

On CPPPCC, the 2 is in the "a" position (first progression sequence), leaving the "b" and "c" for the master key, giving it authority over one bitting per block in each block of a single page of the progression pattern.

On PCCPCP and PCCPPC, the 2 is once again in the second position of the progressed positions, making the master a Horizontal Master Key ("a" and "c") using the first and third positions.

On PCPPCC and PPCPCC, the 2 is in the "a" position (first progression sequence), leaving the "b" and "c" for the master key, giving it authority over one bitting per block in each block of a single page of the progression pattern.

There will be exactly ten progression patterns for each master key of this level. That is because the one position that restricts it can only be a P or a C, so in half of the 20 progression patterns for Holding 3 constant out of 6, it will be P and in the other half it will be a C.

We have just covered all the possible incidental master keys for the Holding 3 constant of 6 positions, and you should now realize it is not as difficult as it seemed. Want four such incidental masters of 96 each? Use the patterns of CCPPCP, CCPPPC, CPCPCP, CPCPPC, PCCPCP and PCCPPC with a 2, a 4, a 6 and an 8. Each will be a Horizontal Master Key and will not share any bittings.

That is because one is the "first" horizontal master, the next is a "second", the next is a "third" and the next is a "fourth", so there will be no overlap using progression patterns where the master is a Horizontal Master Key.

Want four more that will not interchange? Use the Horizontal Master Keys from the inverse patterns...PPCCPC, PPCCCP, PCPCPC, PCPCCP, CPPCPC and CPPCCP. That gives you four more master keys operating 96 change keys each that do not interchange. You have just divided this system into eight totally separate master keys under the Grand Master Key, making it a three level system.

It was not really that hard, once you understand how to separate them, was it?

**Reviewing what we have learned, you will see a few things about the Masters if you look:**

**1) SIZE:** Although there will be some intertwining of operational control between Line Masters and Block Masters, and some between Vertical and Horizontal Masters, the keys will never "JUMP" a level. The bitting for a Block Master, for example, will never be the same as that for a Vertical or Horizontal Master.

**2) POSITION:** When a Block Master operates a full block, the position of that block will determine the position of all the other keys it operates. For example, if it operates the full block in the first position on a page, all the other keys it operates will be "firsts" also: first block on other pages, first line across the top of a page, or first combination in top of a block in each of the blocks in a Vertical Column.

If it controlled the second Block on a page, it would control the second block on other pages, second line across the top of a page or pages, and second combination in each block in a Vertical Column.

Similarly, if a Vertical Master operates the first column on a page, it will also control the first column on other pages, will be the first Horizontal Master on other pages, and will control the first line across the top of the Blocks on each page.

How do we determine which combinations are controlled by a particular Master? Of course, it's obvious when it is one of the control groups where we have the master calculated. If the master has the same bitting, it will control the group.

But what about the groups such as the "second combination in each block of a Vertical Column"? How do we spot these?

It is actually very easy. When looking for the control groups of the various Block Masters or Line Masters in a "Progressing 2" progression pattern, each Block Master will operate every key that has the same bitting as it has in its SECOND PROGRESSION SEQUENCE.

When looking for the control group of a Block Master or Line Master in a "Progressing 3" progression pattern, it will control all the keys that have the same cut as it has in the SECOND & THIRD PROGRESSION SEQUENCE.

When looking for the control groups of the various Horizontal and Vertical Masters in a "Progressing 3" progression pattern, it will control all the keys that have the same cut as it has in its third progression sequence.

When looking for the control groups of the various Vertical or Horizontal Masters in a "Progressing 4" progression pattern, it will control all the keys that have the same cut it has in the 3rd and 4th PROGRESSION SEQUENCES.

When looking for the control groups of the various Page Masters in a "Progressing 4" progression pattern, it will control all the keys that have the same cut it has in its FOURTH PROGRESSION SEQUENCE.

When looking for the control groups of the various Block or Line Masters in a "Progressing 4" progression pattern, it will control all the combinations that have the same cut it has in its 2nd, 3rd, & 4th PROGRESSION SEQUENCES.

When looking for the control groups of a Base Master or Group Master, in a "Progressing 5" progression pattern, it will control all the keys with the same cut it has in its 5th PROGRESSION SEQUENCE.

In a "Progressing 5" progression pattern, the Page master operates all with the same cut in the 4th & 5th SEQUENCES. The Vertical or Horizontal Master operates all the combinations with the same cut it has in its 3rd, 4th, & 5th SEQUENCES. The Block or Line Master operates all with the same cut it has in its 2nd, 3rd, 4th, & 5th PROGRESSION SEQUENCES.

**For all of these, look for the cut that LIMITS its operation. That is the cut that defines its operational authority.**

If we were to progress a Holding two positions Constant and Progressing four positions, our TMK becomes the equivalent of a Group Master and each pattern yields four pages of charts. We can break up the key into the master key groupings we already saw, and also into the equivalent of the Page Master Key.

However, our Page Master Key uses he TMK bittings from progression sequences one, two and three ("a", "b" and "c") and there will also be an equivalent key with positions "b", "c" and "d" (second, third and fourth positions) and "a", "b" and "d" (first, second and fourth) and "a", "c" and "d" (first, third and fourth).

As we already saw in the chapter on parallel master keys, the "abd" parallel master operates all the bittings in one column of every page in a group. The "acd" parallel master key operates all the bittings in one block per column on every column in a group. The "bcd" parallel master key operates one bitting per block in every block in the group.

Notice how everything really fits together? Understanding how incidental master keys fir into the capabilities of an RCM system is a strong and valuable tool in the master keying locksmith's toolkit, and is one that few have previously used.

Use it only if you fully understand it. Used properly it is an amazing tool.

# CHAPTER SEVEN:

# THE XMC VARIANT OF THE RCM

Another way exists to assign a limited number of master keys that do not overlap each other. According to Lloyd Seliber, a former Schlage employee in the training department and a former associate of A.J. Hoffman, it was developed by Jerome Andrews, modified slightly by A.J. Hoffman and practiced by Schlage lock for several years. However, it was shown as a "trick" and as such did not draw the attention it deserves. In truth, it is an amazingly effective tool, and yet quite simple. This was pointed out to this author in 2005.

The technique begins by splitting the key into "Master" positions and "Change Key" positions. For a six pin key, this might typically be three positions for Change Keys and two for Master Keys. The sixth position is Held Constant. Once the system is in place, it can be decided, for future expansion needs, whether to consider that position one of the Master positions, and progress those positions as a Holding 1 Constant out of three on the six pin key, or to expand the number of Change Keys by rotating it as a Holding one out of four on a six pin key with the Change Key positions.

If you have a seven pin key, you have even more possibilities and could even use more than one position as the X position to be held constant and then associated with either Masters (M) or Change Keys (C).

On a six pin you are starting out with 16 (or 25 with a five cut KBA such as Single Increment) Master Keys under the TMK, which can be expanded by applying Rotating Constant to the Master positions to three times that many, or 54 (75) Master Key Levels.

On a six pin key you are starting out with 64 (125) Change Keys per Master Key, with the capability to expand that to 256 (625) by applying Rotating Constant to the Change Key positions to four times that many.

The concept starts out by deciding how to divide the key between change key positions and master key positions. Then at least one position is reserved to be held Constant in the original progression pattern. Because we have Change Key Progression positions, Master Key Progression positions and a position or positions held Constant, the P and C we used as a standard in Rotating Constant (formerly H and V when it was also known as Hold and Vary) will not suffice, so the symbols X, M and C are used. X is used for a Held Constant position, M is for chambers progressed for Master Keys and C (Watch out for this one! It is not C for Constant in this variation.) is for Change Key Progression positions.

To better understand the concept of key division, we should examine the Keying System Specification (KSS), which is somewhat the basis for it. In order to properly discuss the Keying System Specification, we need to briefly explain the Key Numbering systems and the Key Bitting Array (KBA).

> Most lock manufacturers have standardized their Key numbering systems to one of three possibilities. The first two are that they use relatively small increments and obtain ten possible depths of cut by doing so, which they either number from zero to nine (where zero represents the most possible metal remaining on the key after a cut has been made and nine represents a cut leaving the least metal remaining on the key) or from one to zero (where one represents the most possible metal remaining on the key after a cut has been made and zero represents a cut leaving the least metal remaining on the key, often called a "ten" cut) as shown in the following illustration:

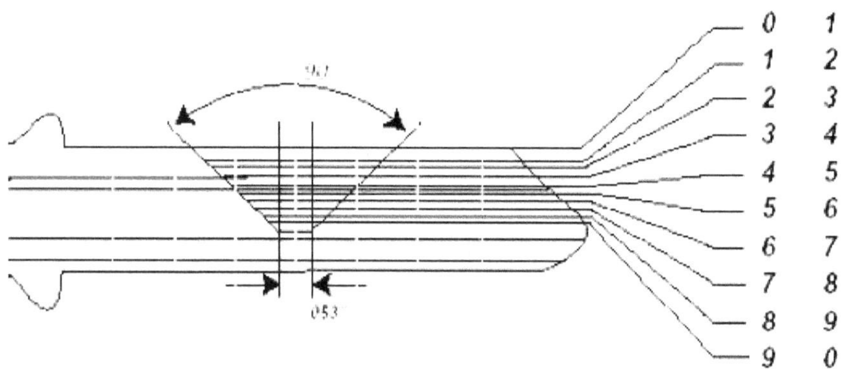


Base **4**

When Master Keying these locks, the typical KBA will have four change key cut options for each position, and everything regarding their progression will be in groups of fours.

The third Key Numbering system is typically from One to Six, (although at least one manufacturer numbers it from Zero to Five) as shown in the following illustration:

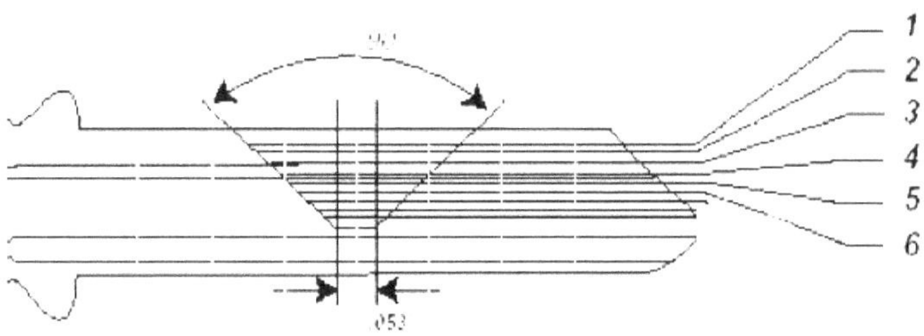


Base **5**

When Master Keying these locks, the typical KBA will have five change key cut options for each position, and everything regarding their progression will be in groups of fives.

If we are using a Standard Progression Format chart (as shown earlier) to lay out the possible cuts for the system, the 0-9 or 1-0 would break up into four bittings per block, four blocks per column, four columns per page, four pages per group, four groups per section, four sections per six pin system and four six pin systems per seven pin system. For this reason, it is sometimes called a Base Four system.

If we are using a Standard Progression Format chart (as shown earlier) to lay out the possible cuts for the system, the 1-6 or 0-5 would break up into five bittings per block, five blocks per column, five columns per page, five pages per group, five groups per section, five sections per six pin system and five six pin systems per seven pin system. For this reason, it is sometimes called a Base Five system.

### → A word about "Base 4" and "Base 5" ←

Sometimes, individuals who believe they understand mathematics object to the terms "Base Four" and "Base Five" using the argument that in a "true" Base four numbering system the numbers are 0, 1, 2 and 3 only and in a "true" Base Five, they are only 0, 1, 2, 3 and 4. What these individuals miss is that numbers do not really exist. They are merely symbols and other symbols may be substituted for them. We could, for example, use A. B. C and D or W, X, Y and Z for a Base four system. In our example we use other numbers, the "value" of which may vary from system to system, but which are merely placeholders. These terms have been in common usage for over thirty years, and master keying locksmiths generally understand the meaning, which is all we really care about when we use the terms.

---

When we look at the system in the view of the KSS, we will be looking at dividing the key based on the levels of master keying in the system. Let's begin by examining a Grand Master Key System (Three Level) and then continue to a four level and a five level. As we look at it for a "Base Four" system, everything will be either 1 (for the top master) or a multiple of four based on the number of chambers involved. As we look at it for a "Base Five" system, everything will be either 1 (for the top master) or a multiple of five based on the number of chambers involved.

Two chambers will yield 16 or 25 bitting combinations
Three chambers will yield 64 or 125 bitting combinations
Four chambers will yield 256 or 625 bitting combinations
Five chambers will yield 1,024 or 3,125 bitting combinations
Six chambers will yield 4,096 or 15,625 bitting combinations
Seven chambers will yield 16,384 or 78,125 bitting combinations

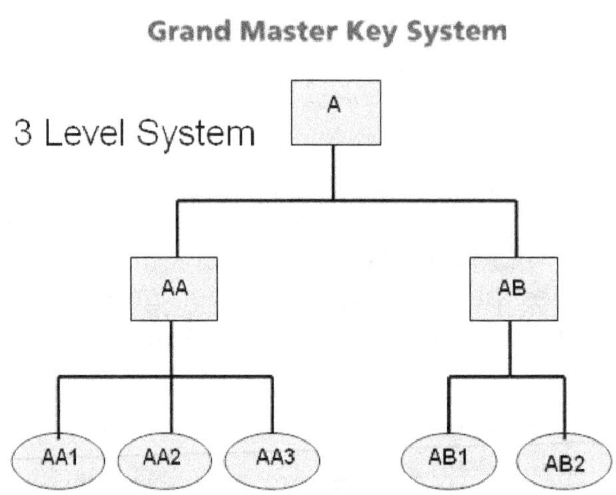

## Grand Master Key Systems

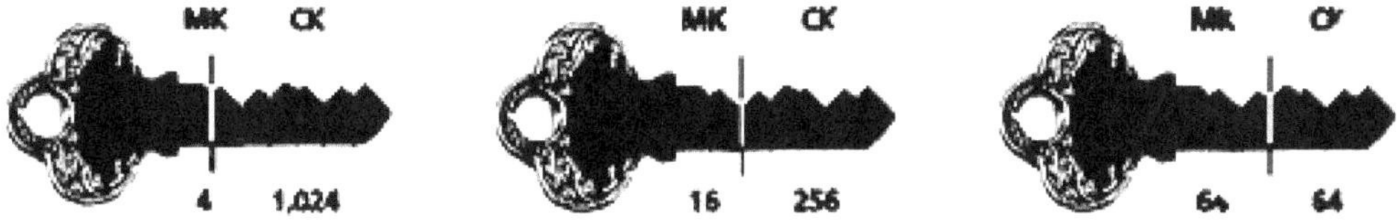


*Grand Master Key Systems*
*6-pin*

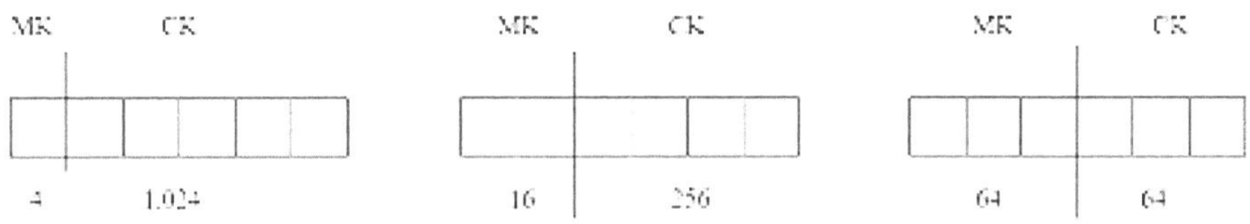


Note: each square represents one pin chamber of the cylinder/key

The charts above represent 1-4-1024; 1-16-256; or 1-64-64.

These are the logical three level groupings for Base 4.

## Grand Master Key Systems

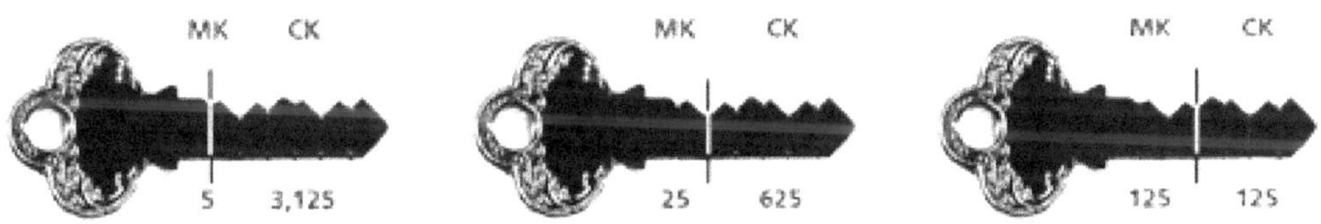


*Grand Master Key Systems*
*6-pin*

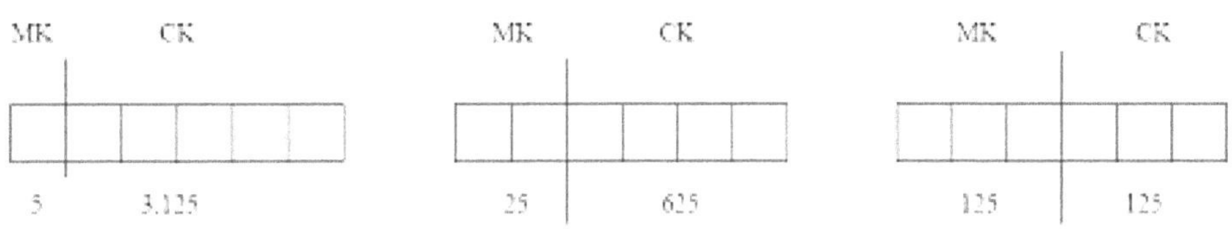


Note: each square represents one pin chamber of the cylinder/key

The charts above represent 1-5-3125; 1-25-625; or 1-125-125

These are the logical three level groupings for Base 5.

## Great Grand Master Key System

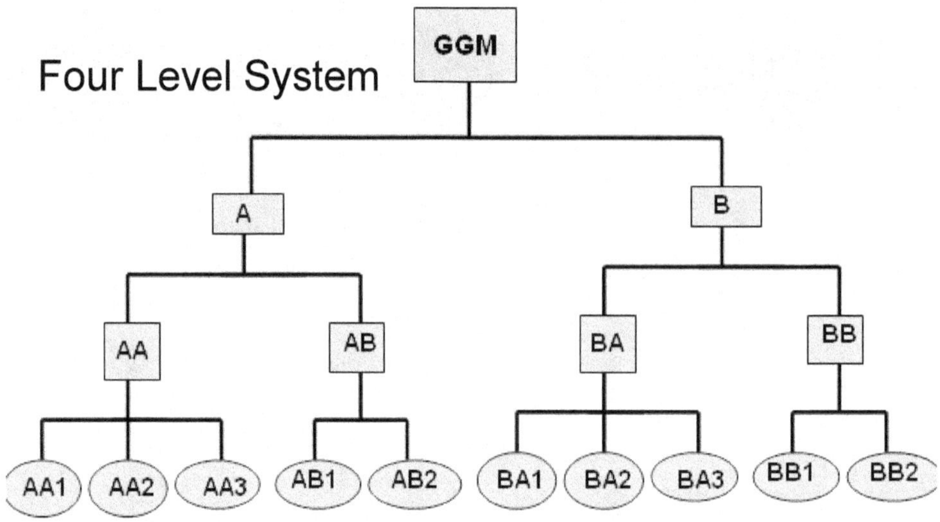


## Great Grand Master Key Systems

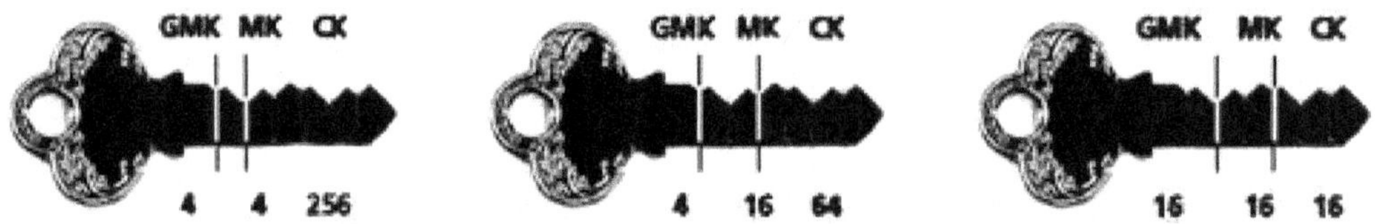


*Great Grand Master Key Systems*
*6-pin*

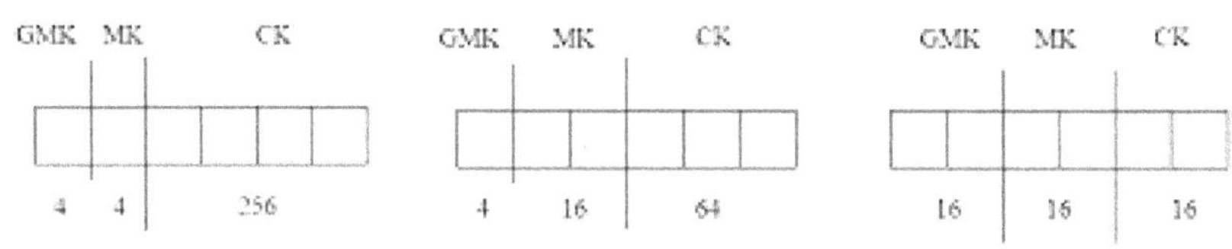


Note: each square represents one pin chamber of the cylinder/key

The charts above represent 1-4-4-256; 1-4-16-64; or 1-16-16-16.

These are the logical four level groupings for Base 4.

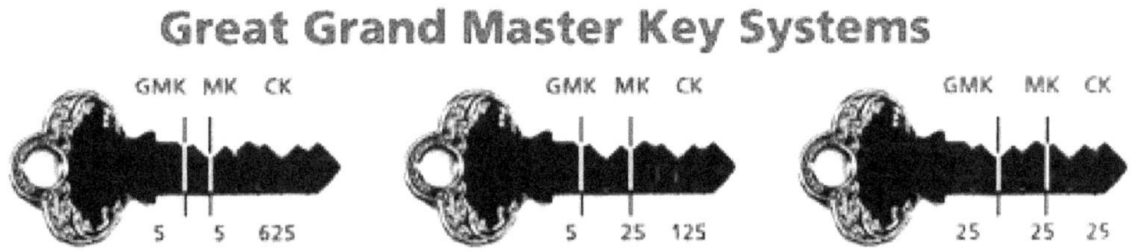


*Great Grand Master Key Systems*
*6-pin*

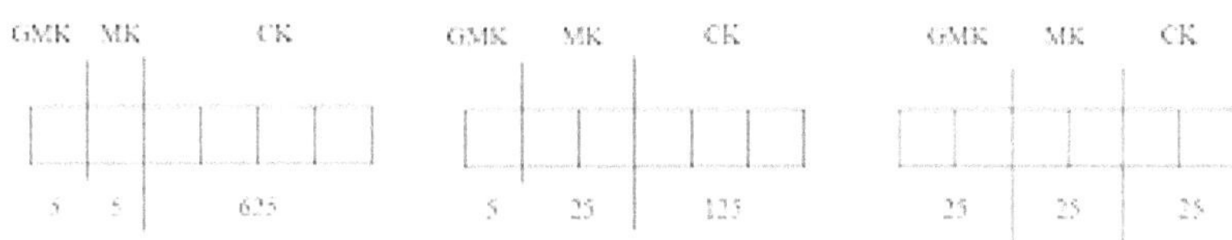


Note: each square represents one pin chamber of the cylinder/key

The charts above represent 1-5-5-625; 1-5-25-125; or 1-25-25-25.
These are the logical four level groupings for Base 5.

The KSS is a useful design tool and really useful when laying out system capabilities. For example, let's say that your original assessment of needs requires 1-30-625. At a glance you could see that it could not be broken up in that manner on a single keyway six pin key.

A good example of exceeding the limitations can be shown by a simple sounding system:

Let's presume that for our customer's Base 4 product's master key system, we needed five masters that could operate 65 combinations each, a grand master to operate all of those, a master that could operate four combinations not under the grand master, and an overall master key.

Despite the fact that it is a big system, it seems fairly easy to lay out, does it not?

One way to determine this would be to examine the level of keying needed by each of the keys.

In a Base Four product:

- A Block or line master can only operate a maximum of four keys.
- A Vertical or Horizontal Master can operate a maximum of 16 keys, or 4 blocks or lines
- A Page Master can operate a maximum of 64 combinations, or 4 columns or lines, or 16 Blocks or lines
- A Group Master can operate a maximum of 256 combinations or 4 pages.
- A Section Master can operate a maximum of 1024 combinations or 4 groups or 16 pages
- A six pin master can operate a maximum of 4,096 combinations or 4 sections, or sixteen groups or 64 pages

Based on that, each of the five sets of 65 keys is too big for a page master, so we must assign a Group master to each. Next we need a key capable of operating these, so we go up a level to the section master...but wait! A section master can only operate FOUR groups, and we have five! That means that it takes the six pin master to act as the grand Master for the five groups. That leaves us no place to pull the four not under the Grand Master and no place to get an overall master for the whole system!

This took a lot of thought when using only the SPF terms to assign, but if we used the KSS, think of how much quicker you would have discovered it, because trying to divide the key (I find the above illustrations VERY helpful) you quickly realize that the grand master completely uses the capabilities of the entire key, leaving no place to pull the extra four combinations and the overall master key from. It doesn't require you to think in terms of exact levels of keying for each.

The difference in using the XMC variant is that the positions indicated on our key divisions under the KSS as being reserved for the "GMK" will instead be reserved as "X" positions, which are treated as Constant positions for this progression, but which may later be progressed using RCM (Rotating Constant Method) by rotating them with either the "Master" positions or with the "Change Key" positions as needed for the system to grow, or, if there are ore than one "GMK" position, one might be expanded using RCM with the "Master" positions, and the other using RCM with the "Change Key" positions.

**THE SAMPLE CHART ON THE NEXT PAGE SHOULD SHOW THIS FAIRLY CLEARLY:**

| | | | | | |
|---|---|---|---|---|---|
| | | | | | |
| | | | | | |
| | | | | | |
| | | | | | |
| | | | | | |

BASE FIELD
(KBA)
KEY BITTING ARRAY

**PAGE MASTER** ____________________

PAGE
# __________
OF __________
PAGES

⇦ SEQUENCE OF PROGRESSION

| C | C | C | M | M | X |
|---|---|---|---|---|---|
| 1 | 4 | 5 | 4 | 6 | 5 |
| 1 | 4 | 6 | 4 | 6 | 5 |
| 1 | 4 | 1 | 4 | 6 | 5 |
| 1 | 4 | 3 | 4 | 6 | 5 |
| 1 | 4 | 4 | 4 | 6 | 5 |
| 1 | 5 | 5 | 4 | 6 | 5 |
| 1 | 5 | 6 | 4 | 6 | 5 |
| 1 | 5 | 1 | 4 | 6 | 5 |
| 1 | 5 | 3 | 4 | 6 | 5 |
| 1 | 5 | 4 | 4 | 6 | 5 |
| 1 | 3 | 5 | 4 | 6 | 5 |
| 1 | 3 | 6 | 4 | 6 | 5 |
| 1 | 3 | 1 | 4 | 6 | 5 |
| 1 | 3 | 3 | 4 | 6 | 5 |
| 1 | 3 | 4 | 4 | 6 | 5 |
| 1 | 2 | 5 | 4 | 6 | 5 |
| 1 | 2 | 6 | 4 | 6 | 5 |
| 1 | 2 | 1 | 4 | 6 | 5 |
| 1 | 2 | 3 | 4 | 6 | 5 |
| 1 | 2 | 4 | 4 | 6 | 5 |

| C | C | C | M | M | X |
|---|---|---|---|---|---|
| 2 | 4 | 5 | 4 | 6 | 5 |
| 2 | 4 | 6 | 4 | 6 | 5 |
| 2 | 4 | 1 | 4 | 6 | 5 |
| 2 | 4 | 3 | 4 | 6 | 5 |
| 2 | 4 | 4 | 4 | 6 | 5 |
| 2 | 5 | 5 | 4 | 6 | 5 |
| 2 | 5 | 6 | 4 | 6 | 5 |
| 2 | 5 | 1 | 4 | 6 | 5 |
| 2 | 5 | 3 | 4 | 6 | 5 |
| 2 | 5 | 4 | 4 | 6 | 5 |
| 2 | 3 | 5 | 4 | 6 | 5 |
| 2 | 3 | 6 | 4 | 6 | 5 |
| 2 | 3 | 1 | 4 | 6 | 5 |
| 2 | 3 | 3 | 4 | 6 | 5 |
| 2 | 3 | 4 | 4 | 6 | 5 |
| 2 | 2 | 5 | 4 | 6 | 5 |
| 2 | 2 | 6 | 4 | 6 | 5 |
| 2 | 2 | 1 | 4 | 6 | 5 |
| 2 | 2 | 3 | 4 | 6 | 5 |
| 2 | 2 | 4 | 4 | 6 | 5 |

| C | C | C | M | M | X |
|---|---|---|---|---|---|
| 3 | 4 | 5 | 4 | 6 | 5 |
| 3 | 4 | 6 | 4 | 6 | 5 |
| 3 | 4 | 1 | 4 | 6 | 5 |
| 3 | 4 | 3 | 4 | 6 | 5 |
| 3 | 4 | 4 | 4 | 6 | 5 |
| 3 | 5 | 5 | 4 | 6 | 5 |
| 3 | 5 | 6 | 4 | 6 | 5 |
| 3 | 5 | 1 | 4 | 6 | 5 |
| 3 | 5 | 3 | 4 | 6 | 5 |
| 3 | 5 | 4 | 4 | 6 | 5 |
| 3 | 3 | 5 | 4 | 6 | 5 |
| 3 | 3 | 6 | 4 | 6 | 5 |
| 3 | 3 | 1 | 4 | 6 | 5 |
| 3 | 3 | 3 | 4 | 6 | 5 |
| 3 | 3 | 4 | 4 | 6 | 5 |
| 3 | 2 | 5 | 4 | 6 | 5 |
| 3 | 2 | 6 | 4 | 6 | 5 |
| 3 | 2 | 1 | 4 | 6 | 5 |
| 3 | 2 | 3 | 4 | 6 | 5 |
| 3 | 2 | 4 | 4 | 6 | 5 |

| C | C | C | M | M | X |
|---|---|---|---|---|---|
| 5 | 4 | 5 | 4 | 6 | 5 |
| 5 | 4 | 6 | 4 | 6 | 5 |
| 5 | 4 | 1 | 4 | 6 | 5 |
| 5 | 4 | 3 | 4 | 6 | 5 |
| 5 | 4 | 4 | 4 | 6 | 5 |
| 5 | 5 | 5 | 4 | 6 | 5 |
| 5 | 5 | 6 | 4 | 6 | 5 |
| 5 | 5 | 1 | 4 | 6 | 5 |
| 5 | 5 | 3 | 4 | 6 | 5 |
| 5 | 5 | 4 | 4 | 6 | 5 |
| 5 | 3 | 5 | 4 | 6 | 5 |
| 5 | 3 | 6 | 4 | 6 | 5 |
| 5 | 3 | 1 | 4 | 6 | 5 |
| 5 | 3 | 3 | 4 | 6 | 5 |
| 5 | 3 | 4 | 4 | 6 | 5 |
| 5 | 2 | 5 | 4 | 6 | 5 |
| 5 | 2 | 6 | 4 | 6 | 5 |
| 5 | 2 | 1 | 4 | 6 | 5 |
| 5 | 2 | 3 | 4 | 6 | 5 |
| 5 | 2 | 4 | 4 | 6 | 5 |

6

| C | C | C | M | M | X |
|---|---|---|---|---|---|
| 6 | 4 | 5 | 4 | 6 | 5 |
| 6 | 4 | 6 | 4 | 6 | 5 |
| 6 | 4 | 1 | 4 | 6 | 5 |
| 6 | 4 | 3 | 4 | 6 | 5 |
| 6 | 4 | 4 | 4 | 6 | 5 |
| 6 | 5 | 5 | 4 | 6 | 5 |
| 6 | 5 | 6 | 4 | 6 | 5 |
| 6 | 5 | 1 | 4 | 6 | 5 |
| 6 | 5 | 3 | 4 | 6 | 5 |
| 6 | 5 | 4 | 4 | 6 | 5 |
| 6 | 3 | 5 | 4 | 6 | 5 |
| 6 | 3 | 6 | 4 | 6 | 5 |
| 6 | 3 | 1 | 4 | 6 | 5 |
| 6 | 3 | 3 | 4 | 6 | 5 |
| 6 | 3 | 4 | 4 | 6 | 5 |
| 6 | 2 | 5 | 4 | 6 | 5 |
| 6 | 2 | 6 | 4 | 6 | 5 |
| 6 | 2 | 1 | 4 | 6 | 5 |
| 6 | 2 | 3 | 4 | 6 | 5 |
| 6 | 2 | 4 | 4 | 6 | 5 |

This page left intentionally blank.

# CHAPTER EIGHT:

# MASTER KEY SYSTEM DECONSTRUCTION AND SOME DEFENSES AGAINST IT

It is important to know that a system can be violated and that care should be taken to protect it. Let's examine the two most serious threats to the system.

As we have gone through various steps in the master keying process, we always keep coming back to the threats we perceive to be awaiting our systems. Let's examine the two biggest threats to master key systems today:

## The Dayton Method of Decoding a Master Key

**In the September, 1958 issue of Locksmith Ledger, James Dayton, Jr., wrote an article on decoding a master key system once either a master key or change key was known, based on simple progression of the known possibilities for each position.**

**It remained a closely kept secret among top master keying professionals since that time.**

**This method has recently been released to the public through print and television media and on the internet by persons not directly related to locksmithing with whom the secret was shared.** One of the people involved in its release also teaches it to groups of hackers, crackers, amateur lock pickers, and college computer science students, expanding awareness of the technique dramatically.

**As a result of the "outing" of the Dayton Method, System Specifiers and Commercial Locksmiths can no longer simply sell a master key system and walk away from it.**

**At the very least they need to instruct the customer's key issue representatives in the proper procedures to defend against the Dayton Method,** including enforceable and strict lock and key policies, and a quick response to lost, stolen or otherwise misplaced keys. They need to instruct the customer/client in educating their employees on what signs to watch for (such as brass filings or key chips that show a key creation attempt of some sort is under way) which indicate a Dayton Method attack may be in the process of being attempted.

**Institutional Locksmiths must do this and more.** They must be vigilant for any situations that make the method more possible or indications that it might be occurring, and evaluate the usable life of the master key system at any stage.

**And all three groups need to understand budgeting for the master key system's life cycle. (more on this later)**

**The technique is so simple that a youngster can easily grasp its concept, though actually implementing it requires the ability to make "keys to dimension" whether with a hand file or a specialized code cutting machine.**

**Along with understanding the method and its steps comes a few other simple steps to resist the method**.

**The first step is to obtain a key that works somewhere in the system and copy it.**

**Therefore the first step in defense against this method** is to restrict as much as possible the keys that are issued, and attempt to issue them only to those with something to lose if the area were compromised.

**The second step is to try the key in its intended lock cylinder** and make certain the copy works. Then the intruder chooses a position where the key has a high cut on it, and cuts it down one depth at a time (or two if the intruder is confident a two increment drop was used in the master key system) and tries it in the lock, until it finds the second position where it operates. This is believed to be the master key cut depth for that position. They then move to another position and try the procedure again.

**Common areas such as lunchrooms, copier rooms, restrooms, etc., are typically not considered high security areas**. They are also areas where an outsider might easily obtain a key on a temporary basis, make a copy for it and return the original key, and attempt the method. Restrooms in particular are vulnerable because their placement usually puts them in a condition of being somewhat out of the public view where the would-be intruder can work unobserved. Leaving pins out of these cylinders can be an excellent technique to make it more difficult for the intruder to accomplish his goal of decoding the system from that key they were issued for that area. This is a second defense method.

**The use of keys that are resistant to copying or the obtaining of blanks to match is another simple step that resists the method.** The person obtaining a key for the restroom now finds that not only would the method not work as well, but they will find it difficult to obtain and cut the necessary keys called for by the method. This is a third defense.

**The act of trying the key after each successive cut means that the intruder would like to work unobserved for a period of time at a particular door**. Because it is unlikely that they will be able to do so uninterrupted, they are likely to have to periodically move away from the area. But the key filings or key chips created by the method are likely to remain behind as evidence that an attack has been or is being tried. Vigilance against that observable evidence is therefore another defense against the technique.

**Using non-progressed positions is another defense tactic to slow them down.** If the intruder gets to the lowest of the possible cuts and has not found the second operating position, they must then start at the highest possible cut for that position, and continue back to the original cut. If they get to the original cut and have not found the second operating depth, either they missed it or it is a non-progressed position. But what if instead of obtaining a key through legitimate channels the person simply stole or found the key? **Now the first question is, does the key stamping telegraph information and if so, how much? It is less common today than it used to be but at one time it was quite common for a key to be stamped with its operating location (such a Room 213). In such a case, the intruder can quickly determine where to try the key.**

**If instead it is stamped with SKC (Standard Key Coding symbols) it slows them down a little**, but observing other keys as they are used can identify the general area in many cases. Of course, if the SKC is also stamped on the face of the cylinder, it is almost as easy as if it simply listed its location.

**Blind coding on the keys and the use of CKC on the other hand, conveys little or no information,** so another defense against the Dayton Method is to use Blind Coding.

**The next defense against the method is to deactivate keys as quickly as possible when they are not in the proper issue's possession.**

**At the time of issue, it should be explained in no uncertain terms that it is the issuee's responsibility to report its loss promptly. That should also be agreed to in writing, and should be supported by enforceable penalties for failing to do so.**

**Once the report is given, a re-key of the affected area(s) should be automatic.**

**That may seem a little expensive and harsh at first look,** but if you have properly laid out the procedures, the money to do so should already have been budgeted. We will get into that much more very shortly.

**Many companies try to punish the person who reports it in order to "discourage irresponsibility".** That may have been a workable solution at one time, but with the threat of the Dayton Method looming overhead, the cost of re-keying an individual door or small area is much less than the possible threat if it remains in use long enough to allow the Dayton Method attack to succeed.

**And realistically, charging generally has little effective results when it comes to preventing losses. Instead it results in preventing REPORTED losses.**

**It should be pointed out here that a master keyed apartment complex is particularly vulnerable to the Dayton Method** because the renter has unlimited access to a working key and a cylinder to try it in, with no one to observe anything unusual.

**Student rooms in University residence halls are another area where the possibility is quite high.** These should be either on small master key groups so as to at least limit the area of vulnerability or should be on keys that resist duplication.

# Disassembly Attack

**The amount of skill required to access and disassemble a lock cylinder varies with the product type and grade.** However, some (usually grade three residential locks) require only a minimal amount of knowledge and skills to do this. These are exceptionally vulnerable to decoding and reassembly and pose a legitimate threat to system security. Without naming brand names or specific product types here, there are cylinders that can be "popped out" with a small screwdriver and disassembled by removing a snap-on retainer cap. Others may have a brass slide that must be removed to gain access to the chambers. Some cylinders are simply screwed into the door with a retaining screw from the door edge to prevent their being unscrewed. Rarely is it all that difficult to gain access to the cylinder itself, or once access has been gained, to remove and measure the cylinder pins.

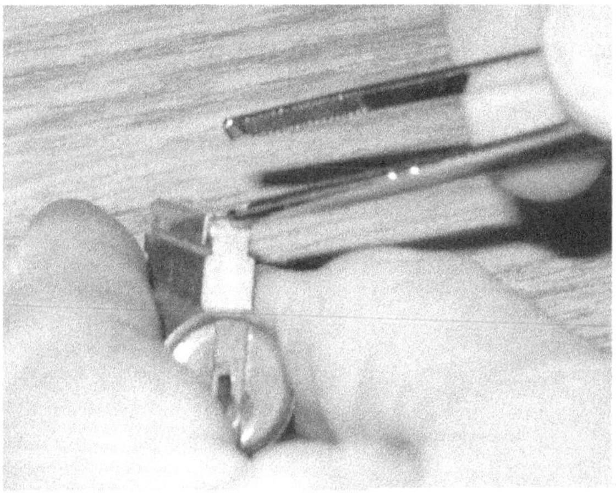

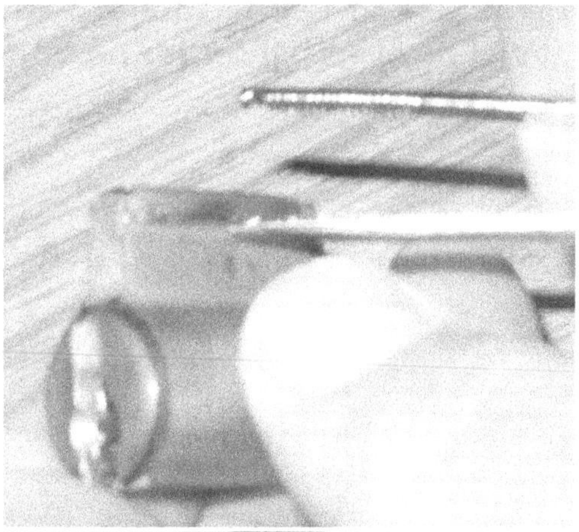

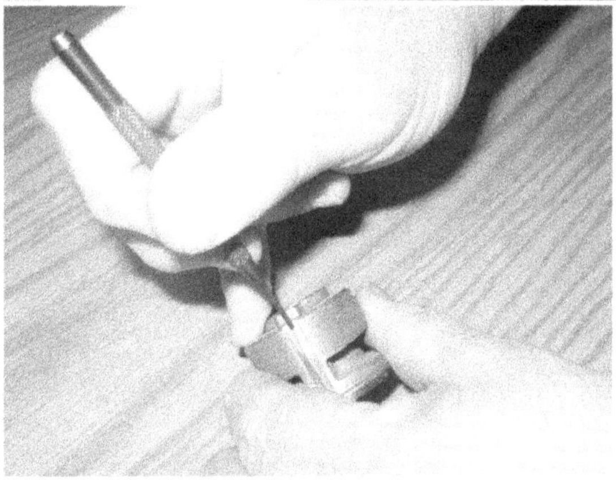

**Other types of cylinders (such as Master Ring cylinders in grade one locksets or interchangeable core products) may be difficult to remove from the lockset, difficult to disassemble, and slightly more difficult to decode once they have been disassembled.**

**Patent Protected keys are keys which are difficult, though rarely impossible, to obtain except through legitimate channels.** This means that even if a would-be intruder wishes to disassemble the lock cylinder, he or she will be unable to easily fit a key to operate it.

**Remember that this protection is not absolute.** There are key machines to create duplicate key blanks that are readily available to locksmiths, and with the advent of the internet in its current state, may be available to the public as well in many cases. Even original blanks might be available on the internet, regardless of manufacturer contracts to the contrary.

**This, by the way, also offers a great degree of protection against the Dayton Method. If the intruder would find it difficult to obtain one blank key to use in fitting a key by the Disassembly Method, then it will be even harder for he or she to obtain six or more to use the Dayton method.**

# CHAPTER NINE:

# UNDERSTANDING THE LIFE CYCLE OF A MASTER KEY SYSTEM

## A Master Key System is a mechanical device.

### Like any mechanical device its life cycle has five stages:

- Installation stage
- Adjustment stage
- Maintenance stage
- Repair stage
- Replacement stage

**By analyzing and budgeting for these stages, emergency or crisis spending can be avoided.**

**Perhaps more importantly, poorly made decisions based on unfunded crisis expenses can be avoided.**

**Understanding the life cycle and sharing it with your client or customer can be an important step in determining how effectively you have performed your duties in master key system layout, and in how prepared you are in advance of crisis situations that will inevitably occur.**

# Installation Stage

**This includes the actual design and specification as well as the intended application of the system.**

- **First determine if a master key system is justified and needed.** An apartment complex might be better served by a keybox accessible to the maintenance man with individual keys in it, and no master key in existence, to give better security and protect against the Dayton Method, to which they are exceptionally vulnerable.

- **Choose the appropriate methods, formats and presentation style for the master key system and its lists.**

- **Consider patent protected keys or high security cylinders or cores.**

- **"Walk the system" and lay out the anticipated needs using SKC symbols and a user grid chart.**

- **Educate those in charge of key distribution about standard procedures and why they are important.**

# Adjustment Stage

**Sometimes called the "fine-tuning stage",** this is the time to correct mistakes in layout, handle added or removed rooms, changed room numbers and changed patterns of usage or similar changes that might have occurred since original walk-through.

**These are virtually inevitable in any system**, but the longer the period between layout and the application of the cylinder or cores to the doors, the greater the number of changes that are likely to have occurred.

**This should be a short stage of only a few weeks or months at most, and frequently only a few days**

# Maintenance Stage

**This should be the longest stage.** During it, key policies and key issue, tracking and return should all be enforced.

**As keys for single areas are lost, stolen or otherwise not accounted for, the areas must be re-keyed promptly.**

- Custodial and maintenance people must be reminded frequently to be careful about their keys and not leave them laying about or hanging in a door.
- Employees and other clients such as students in a University must be reminded periodically not to attempt to duplicate the keys and to report lost keys promptly.

**Remember: A key "in the wild" can cost a lot more than the replacement or re-keying even if the institution has to "eat" the charges.**

The lock and key policies and enforcement will determine how long this stage lasts.

- In a hospital, it is typical for this to last 15 years or more.
- In a university, if custodial and maintenance sign out building keys as needed instead of being issued master keys for several areas, and student lock cores or cylinders are changed when keys are lost, the stage typically lasts 12 to 15 years (up to double that for patent protected keys or high security cores or cylinders).
- Issuing master levels to custodial and maintenance personnel reduces it to 8 to 10 years.
- Failure to change the affected locks reduces it to six.

**Whatever this stage is, it will usually be fairly constant in length unless policies change.** Therefore, knowing the history of key loss and master key loss from previous systems can give a relatively valid picture of how long this stage will be unless changes occur in policies.

# Repair Stage

**This is your warning period.** People are becoming comfortable with the system and making exceptions to the rules. As a result master keys of low levels are being mis-placed, lost or stolen, and individual keys are being replaced more frequently. If one or the other appears to be occurring without the other, it is a warning that losses are possibly not being reported.

**During this stage it is necessary to re-key smaller areas throughout the institution.**

**The system is beginning to break down.** Usually this stage is no more than two to three years, and it can be as short as six months, depending on lock and key policies and their enforcement.

# Replacement Stage

**The system has been compromised or it is likely that the system has been compromised.**

**It is time to replace the system with a new one**. If the maintenance stage was long, it is wise to consider new cores or cylinders. When you do, be aware that patent protected keyways or restricted keyways can potentially double the lifetime of the system.

**If you have anticipated the life cycle correctly, the person or group paying for the replacement has been budgeting as time went by, and is prepared for the situation.**

**If not, it is now a crisis and everyone is scrambling to find the money, put the blame on someone else, or find an excuse not to do it.** That person or group now looks incompetent to those above them, and neither they nor those above them think very highly of you.

# CHAPTER TEN: MASTER KEYING TIPS AND QUICK PINNING CHARTS FOR SOME SPECIAL PRODUCTS:

The A2 SFIC (Small Format Interchangeable Core) products have a numbering system from 0 to 9 and a stack height of 23 in every chamber. The control lug is 10 increments in thickness so 10 is the Control Additive. The use of a nine cut in an even parity chamber position (and elimination of the 8 and 0 from usage in that position) gives a three increment difference between the Control Key and any other change keys. Progress the chart as you normally would. There is no MACS (Maximum Adjacent Cut Specification) violation capability so you will lose no key bittings to the MACS. Instead, simply cross off any bittings containing the 8 depth in the position you chose for the 9 cut on the control key. You cross off the zero in that position because it would require a .0125" thick #1 Control Pin (a.k.a. Build-up pin) which does not exist.

To use the following chart, calculate the TMK and Change Key bittings to determine the Bottom Pin and Master Pin in the standard manner. Total them up per position and then trace down the chart to the Control Key cut for that position. Where they intersect is the Control Pin required and to the far right of it is shown the Driver to make the stack height of 23.

**SFIC A2**

| | Total of Masters and Change Keys | | | | | | | | | | |
|---|---|---|---|---|---|---|---|---|---|---|---|
| Control Key Number | 0 | 1 | 2 | 3 | 4 | 5 | 6 | 7 | 8 | 9 | Drivers |
| 0 | 10 | 9 | 8 | 7 | 6 | 5 | 4 | 3 | 2 | | 13 |
| 1 | 11 | 10 | 9 | 8 | 7 | 6 | 5 | 4 | 3 | 2 | 12 |
| 2 | 12 | 11 | 10 | 9 | 8 | 7 | 6 | 5 | 4 | 3 | 11 |
| 3 | 13 | 12 | 11 | 10 | 9 | 8 | 7 | 6 | 5 | 4 | 10 |
| 4 | 14 | 13 | 12 | 11 | 10 | 9 | 8 | 7 | 6 | 5 | 9 |
| 5 | 15 | 14 | 13 | 12 | 11 | 10 | 9 | 8 | 7 | 6 | 8 |
| 6 | 16 | 15 | 14 | 13 | 12 | 11 | 10 | 9 | 8 | 7 | 7 |
| 7 | 17 | 16 | 15 | 14 | 13 | 12 | 11 | 10 | 9 | 8 | 6 |
| 8 | 18 | 17 | 16 | 15 | 14 | 13 | 12 | 11 | 10 | 9 | 5 |
| 9 | 19 | 18 | 17 | 16 | 15 | 14 | 13 | 12 | 11 | 10 | 4 |

It should probably be pointed out this is was the method suggested by Gerry Finch and A.J. Hoffman, and is NOT the way the majority of manufacturers, including BEST, choose the Control Key. Most simply use a bitting from a specific position on the charts and cross off that bitting. Some use "Altered Parity" and choose a Control Key that will not appear on the chart. For example, if the parity of the system was EEOOOE, they might use OOEOEE.

The A3 SFIC (Small Format Interchangeable Core) products have a numbering system from 0 to 6 and a stack height of 16 in every chamber. The control lug is 7 increments in thickness so 7 is the Control Additive. There is no MACS (Maximum Adjacent Cut Specification) violation capability so you will lose no key bittings to the MACS. This system is considered obsolete and new systems should NOT be developed using it. The chart is for usage should you find a hospital of governmental facility still using it.

To use the following chart, calculate the TMK and Change Key bittings to determine the Bottom Pin and Master Pin in the standard manner. Total them up per position and then trace down the chart to the Control Key cut for that position. Where they intersect is the Control Pin required and to the far right of it is shown the Driver to make the stack height of 16.

## SFIC A3

| Control Key | Total of Masters and Change Keys: 0 | 1 | 2 | 3 | 4 | 5 | 6 | Drivers |
|---|---|---|---|---|---|---|---|---|
| 0 | 7 | 6 | 5 | 4 | 3 | 2 | 1 | 9 |
| 1 | 8 | 7 | 6 | 5 | 4 | 3 | 2 | 8 |
| 2 | 9 | 8 | 7 | 6 | 5 | 4 | 3 | 7 |
| 3 | 10 | 9 | 8 | 7 | 6 | 5 | 4 | 6 |
| 4 | 11 | 10 | 9 | 8 | 7 | 6 | 5 | 5 |
| 5 | 12 | 11 | 10 | 9 | 8 | 7 | 6 | 4 |
| 6 | 13 | 12 | 11 | 10 | 9 | 8 | 7 | 3 |

The A4 SFIC (Small Format Interchangeable Core) products have a numbering system from 0 to 5 and a stack height of 14 in every chamber. The control lug is 6 increments in thickness so 6 is the Control Additive. Progress the chart as you normally would using single increment drop. There is no MACS (Maximum Adjacent Cut Specification) violation capability so you will lose no key bittings to the MACS. Then choose one or two positions to cross off the cut depth used by the Control Key in.

To use the following chart, calculate the TMK and Change Key bittings to determine the Bottom Pin and Master Pin in the standard manner. Total them up per position and then trace down the chart to the Control Key cut for that position. Where they intersect is the Control Pin required and to the far right of it is shown the Driver to make the stack height of 14.

**SFIC A4**

**Total of Masters and Change Keys**

| Control Key | 0 | 1 | 2 | 3 | 4 | 5 | Driver |
|---|---|---|---|---|---|---|---|
| 0 | 6 | 5 | 4 | 3 | 2 | 1 | 8 |
| 1 | 7 | 6 | 5 | 4 | 3 | 2 | 7 |
| 2 | 8 | 7 | 6 | 5 | 4 | 3 | 6 |
| 3 | 9 | 8 | 7 | 6 | 5 | 4 | 5 |
| 4 | 10 | 9 | 8 | 7 | 6 | 5 | 4 |
| 5 | 11 | 10 | 9 | 8 | 7 | 6 | 3 |

Sargent Large Format 6300 series cores use a numbering system from 1 to 0 (deep) with a stack height of 20 in chambers 3 and 4 (which have the control lug) and a stack height of 15 in the remaining chambers. The control lug is 8 increments in thickness so 8 is the Control Additive. Progress the chart as you normally would. There is a MACS (Maximum Adjacent Cut Specification) of 7, so bittings with a 1 next to a 9 or 0, or a 2 next to a 0 are violations of the MACS and must be crossed off. (Remember that only the bitting containing the violation is crossed off, so if a particular master key violates it, that key is crossed off, but the keys "under" it are not crossed off unless they also contain the MACS violation. ) Because the Control lug is 8 increments of thickness, some combinations are also unusable as shown in the chart. To use the following chart, calculate the TMK and Change Key bittings to determine the Bottom Pin and Master Pin in the standard manner. Total them up per position and then trace down the chart to the Control Key cut for that position. Where they intersect is the Control Pin required for positions 3 or 4 and to the far right of it is shown the Driver to make the stack height of 20. For positions other than 3 and 4 there is no control pin used, and the drivers across the bottom match the pins you have indicated for the Bottom Pin and Master Pin to achieve the stack height of 15.

**Sargent 6300 Removable Core**

| Control Key | Total of Masters and Change Keys | | | | | | | | | | ▼Driver Chambers 3&4 |
|---|---|---|---|---|---|---|---|---|---|---|---|
| | 1 | 2 | 3 | 4 | 5 | 6 | 7 | 8 | 9 | 0 | |
| 1 | 8 | 7 | 6 | 5 | 4 | 3 | 2 | | | | 11 |
| 2 | 9 | 8 | 7 | 6 | 5 | 4 | 3 | 2 | | | 10 |
| 3 | 10 | 9 | 8 | 7 | 6 | 5 | 4 | 3 | 2 | | 9 |
| 4 | 11 | 10 | 9 | 8 | 7 | 6 | 5 | 4 | 3 | 2 | 8 |
| 5 | 12 | 11 | 10 | 9 | 8 | 7 | 6 | 5 | 4 | 3 | 7 |
| 6 | 13 | 12 | 11 | 10 | 9 | 8 | 7 | 6 | 5 | 4 | 6 |
| 7 | 14 | 13 | 12 | 11 | 10 | 9 | 8 | 7 | 6 | 5 | 5 |
| 8 | | 14 | 13 | 12 | 11 | 10 | 9 | 8 | 7 | 6 | 4 |
| 9 | | | 14 | 13 | 12 | 11 | 10 | 9 | 8 | 7 | 3 |
| 0 | | | | 14 | 13 | 12 | 11 | 10 | 9 | 8 | 2 |
| ▲Driver Chamber 1,2,5 &6 | 14 | 13 | 12 | 11 | 10 | 9 | 8 | 7 | 6 | 5 | |

Corbin-Russwin has a Master Ring cylinder (Jumbo cylinder) where the second shearline is in all positions. They have offered a standard removable core with positions 3,4,5 and 6 which uses drivers to match the control key cuts in those positions and a .247" in the remaining position, and a High security Interlocking Pin core where the control lug is in positions 2 and 3 only. Corbin's primary numbering systems have been 1 to 0 (deep) and 1 to 6 (using single increment drop master keying) Russwin also had a 0 (shallow) to 9 numbering system on its standard cores on some product lines. The following charts cover all of those, but remember that the driver shown is only for control lug positions on their removable and interchangeable cores.

**Corbin / Russwin Removable Core 1-6 and 1-0**

| | Total from Operating Shear Line | | | | | | | | | | |
|---|---|---|---|---|---|---|---|---|---|---|---|
| Control Key | 1 | 2 | 3 | 4 | 5 | 6 | 7 | 8 | 9 | 10 | Drivers |
| 1 | 0 | -1 | -2 | -3 | -4 | -5 | -6 | -7 | -8 | -9 | 1 |
| 2 | +1 | 0 | -1 | -2 | -3 | -4 | -5 | -6 | -7 | -8 | 2 |
| 3 | +2 | +1 | 0 | -1 | -2 | -3 | -4 | -5 | -6 | -7 | 3 |
| 4 | +3 | +2 | +1 | 0 | -1 | -2 | -3 | -4 | -5 | -6 | 4 |
| 5 | +4 | +3 | +2 | +1 | 0 | -1 | -2 | -3 | -4 | -5 | 5 |
| 6 | +5 | +4 | +3 | +2 | +1 | 0 | -1 | -2 | -3 | -4 | 6 |
| 7 | +6 | +5 | +4 | +3 | +2 | +1 | 0 | -1 | -2 | -3 | 7 |
| 8 | +7 | +6 | +5 | +4 | +3 | +2 | +1 | 0 | -1 | -2 | 8 |
| 9 | +8 | +7 | +6 | +5 | +4 | +3 | +2 | +1 | 0 | -1 | 9 |
| 10 | +9 | +8 | +7 | +6 | +5 | +4 | +3 | +2 | +1 | 0 | 10 |

**Non-control positions use .247" drivers in cores and .177" drivers in master ring cylinders.**

## Corbin / Russwin Master Ring

| Lowest Number from Master Ring Shear Line \ Total from Operating Shear Line | 1 | 2 | 3 | 4 | 5 | 6 | 7 | 8 | 9 | 0 |
|---|---|---|---|---|---|---|---|---|---|---|
| 1 | 0 | -1 | -2 | -3 | -4 | -5 | -6 | -7 | -8 | -9 |
| 2 | +1 | 0 | -1 | -2 | -3 | -4 | -5 | -6 | -7 | -8 |
| 3 | +2 | +1 | 0 | -1 | -2 | -3 | -4 | -5 | -6 | -7 |
| 4 | +3 | +2 | +1 | 0 | -1 | -2 | -3 | -4 | -5 | -6 |
| 5 | +4 | +3 | +2 | +1 | 0 | -1 | -2 | -3 | -4 | -5 |
| 6 | +5 | +4 | +3 | +2 | +1 | 0 | -1 | -2 | -3 | -4 |
| 7 | +6 | +5 | +4 | +3 | +2 | +1 | 0 | -1 | -2 | -3 |
| 8 | +7 | +6 | +5 | +4 | +3 | +2 | +1 | 0 | -1 | -2 |
| 9 | +8 | +7 | +6 | +5 | +4 | +3 | +2 | +1 | 0 | -1 |
| 0 | +9 | +8 | +7 | +6 | +5 | +4 | +3 | +2 | +1 | 0 |

**Non-control positions use .177" driver in Master Ring.**

**Corbin / Russwin Removable Core 0 to 9**

| | | Total from Operating Shear Line | | | | | | | | | | |
|---|---|---|---|---|---|---|---|---|---|---|---|---|
| | | 0 | 1 | 2 | 3 | 4 | 5 | 6 | 7 | 8 | 9 | |
| Control Key | 0 | 0 | -1 | -2 | -3 | -4 | -5 | -6 | -7 | -8 | -9 | 0 |
| | 1 | +1 | 0 | -1 | -2 | -3 | -4 | -5 | -6 | -7 | -8 | 1 |
| | 2 | +2 | +1 | 0 | -1 | -2 | -3 | -4 | -5 | -6 | -7 | 2 |
| | 3 | +3 | +2 | +1 | 0 | -1 | -2 | -3 | -4 | -5 | -6 | 3 |
| | 4 | +4 | +3 | +2 | +1 | 0 | -1 | -2 | -3 | -4 | -5 | 4 |
| | 5 | +5 | +4 | +3 | +2 | +1 | 0 | -1 | -2 | -3 | -4 | 5 |
| | 6 | +6 | +5 | +4 | +3 | +2 | +1 | 0 | -1 | -2 | -3 | 6 |
| | 7 | +7 | +6 | +5 | +4 | +3 | +2 | +1 | 0 | -1 | -2 | 7 |
| | 8 | +8 | +7 | +6 | +5 | +4 | +3 | +2 | +1 | 0 | -1 | 8 |
| | 9 | +9 | +8 | +7 | +6 | +5 | +4 | +3 | +2 | +1 | 0 | 9 |

(Right-hand column: Drivers)

**Non-control positions use .247" drivers in cores and .177" drivers in master ring cylinders.**

There are several methods by which the Master Ring cylinder systems might have been progressed. Early systems may have taken advantage of the fact that with one cut (or set of cuts) at one shearline and another at the other shearline, it was possible to use cuts that were only one .014" increment apart without interchange. A minimum of two increments apart was still necessary for any bittings operating at the same shearline as each other, though. Sometimes the TMK would be pinned at the lower shearline and the Change key and area key at the upper shearline, instead of the more common way, in order to use more "plus" pins in calculating the pinning.

HOWEVER, if you are developing a NEW system from scratch, it is suggested that you either use the system 70 depths (for single increment drop) or maintain a two increment minimum drop on all bittings, whether Change Key or Master Key. Remember that no one has made a Master Ring cylinder to fit a leverset and most new systems will eventually have to be brought to ADA specifications. You don't want to have to create a new master key system and trash a system you recently designed simply because you failed to provide for the use of leversets! Your customer/client will not be happy with such a scenario.

Also, the use of jumbo cylinders limits you to one brand of hardware. If a renovation is made and the specifier did not use that type of hardware or used types of hardware from that manufacturer, but not ones that accept the jumbo cylinder, once again your system will be worthless if you did not use wither System 70 depths (that are already two cut depths apart, or .028" increments) or use a minimum of a two increment drop.

The specialty products you have seen in this chapter are common enough in master keyed systems that a lack of knowledge on your part of how to proceed in master keying them could be construed as a general lack of knowledge of master keying, but they are really just another useful tool in the Advanced Master Keying toolkit!

# CHAPTER 11
# Matrix

***A set of shorthand arrays that permit the entire chart to be viewed on as little as one page***

AA1- 830852

**AA2- 832492**

| 4 | 9 | 6 | 0 | 3 | 6 |
|---|---|---|---|---|---|
| 6 | 1 | 8 | 2 | 5 | 8 |
| 8 | 3 | 0 | 4 | 7 | 0 |
| 0 | 5 | 2 | 6 | 9 | 2 |
| 2 | 7 | 4 | 8 | 1 | 4 |
| d | e | a | b | c | f |

| | f |
|---|---|
| 8 | |
| 0 | |
| 2 | A |
| 4 | |

| d \ e | 1 | 3 | 5 | 7 |
|---|---|---|---|---|
| 6 | | | | |
| 8 | | A | | |
| 0 | | | | |
| 2 | | | | |

| a \ bc | 25 | 45 | 65 | 85 | 27 | 47 | 67 | 87 | 29 | 49 | 69 | 89 | 21 | 41 | 61 | 81 |
|---|---|---|---|---|---|---|---|---|---|---|---|---|---|---|---|---|
| 8 | | | | | | | | | | | | | | | | |
| 0 | | | | 1 | | | | | | | | | | | | |
| 2 | | | | | | | | | | 2 | | | | | | |
| 4 | | | | | | | | | | | | | | | | |

In the above chart, bitting 83 came from the de position matrix. The abc matrix gave us the 085 for AA1 and the 249 for the AA2, and the f chart completed the AA assignment with the grand master cut for A being the 2.

Note that as each "pair" (although the a-bc pair is actually three positions and three cuts) is assigned, its relative assignment is marked on the matrix, allowing us to see at a glance what has been used of the system. This is the only presentation style that makes it this easy to find "open combinations for new assignments later on.

Note that it uses standard position sequence letters (small letters to keep them separate from the SKC symbols) where a is the first position to change, and f is the sixth position to change, with a therefore being considered a change key position, and f being considered a master key position.

On charts like this in use by Medeco and other manufacturers, instead of breaking the matrix up by Sequence Of Progression, they directly apply Change Key positions to one chart, Master Key positions to another chart, and Grand Master Key positions on another chart.

But In this book, we can only touch gently upon the technique known as Matrix. Although the general technique can be traced back to Vic Fanberg in the late 1970's, and the topic is shown in virtually all advanced master keying related books, articles and courses including those offered by manufacturers, the indisputable father of the technique as it is used today is Billy B. Edwards, and the reader is directed toward his latest book on Master Keying to really learn the technique and its full potential. At the time of the writing of this book, his latest book is:

# CHAPTER TWELVE: MULTIPLEX KEY SYSTEMS

Up until now, this book has dealt with the parameters of a professional master key system on one keyway. Certain manufacturers, however, utilize a "**Multiplex KEY SYSTEM**" to allow for expansion beyond the normal limits of the system.

A multiplex key system is a group of keyways designed by the manufacturer so that, while each keyway is designed not to interchange with any of the others in the system, there are enough common factors in the design of them, that certain key blanks, designed to utilize these common factors, can pass some or all of these keyways. In other words, there are "master blanks" that, if properly cut, would operate ALL the locks in the system, in spite of the different keyways involved.

If there were nine individual keyways, (not including master blank keyways, since the master keyways are only cut on the blank, not into the plugs) this would expand the possible combinations by nine times. On a lock with ten possible depths,(0 to 9 or 1 to 0) we had 4096 POSSIBLE change keys using a six pin key. Using a multiplex keyway system with nine keyways, we would instead have 36,864 possible change keys! Impressive. right?

Now for the bad news. The security of these systems lies in ONLY the correct keys being cut on the master blanks. However, key cutters in five and dime stores, discount department stores, auto parts stores, (just to name a few--I have even found this happening in Pizza Shoppes, and one Bridal Gown store) do NOT STOCK ALL THE KEYS IN THE GROUP!!! Instead they stock ONLY the master key blank that passes ALL the keyways in the group! No matter WHAT keyway the customer's key comes in, it is duplicated onto the "GRAND MASTER KEY BLANK"!!!

This destroys the security of the whole system! I have even seen a few locksmiths do this!

IN AREAS WHERE LOCKSMITHS AND KEYMAKERS ARE LICENSED, THIS IS A SEVERE CRIME!!! Unfortunately, in most areas, UNLESS THE KEY CAN BE TRACED to a particular locksmith or keymaker AFTER it has been used for a crime, it is usually only UNETHICAL!!! The fact that it is, at the least, unethical, seems to have had no effect on the number of times it occurs!

Some of the multiplex key groups, however, have not become popular enough for these keycutters to bother stocking them, thus, at least temporarily ensuring their security.

The following illustrations will show some of the multiplex key systems currently in use and the way the blanks are designed. Remember that typically only the top row is an actual keyway for the lock cylinder plugs.

The "MASTER" keyways are KEY BLANKS ONLY--not locks!

By the way, the list glossary refers to **multiplex keyway systems** as "**multiplex**", and systems not capable of this as "**simplex**". At the time of this writing the term SIMPLEX has yet to be in common usage. As said throughout the book, learn and use all the applicable terms. It will benefit you.

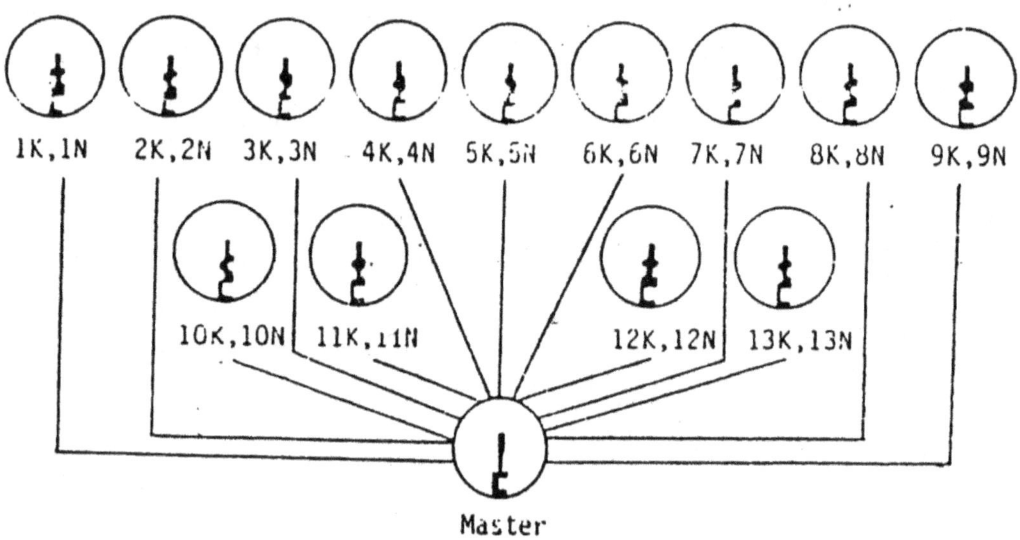


RUSSWIN
980 SERIES
960 SERIES

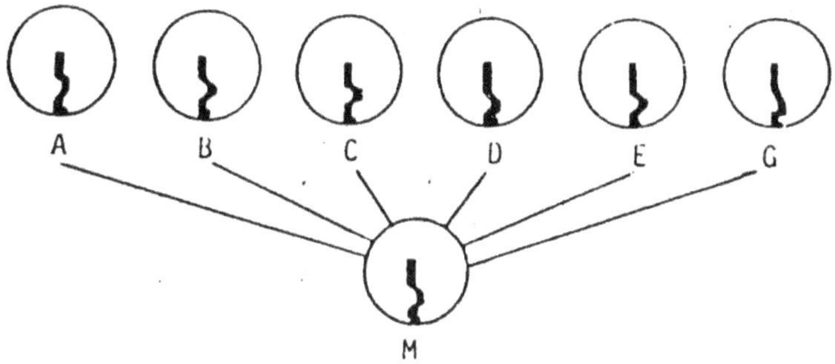


BEST A-M
FALCON
IC

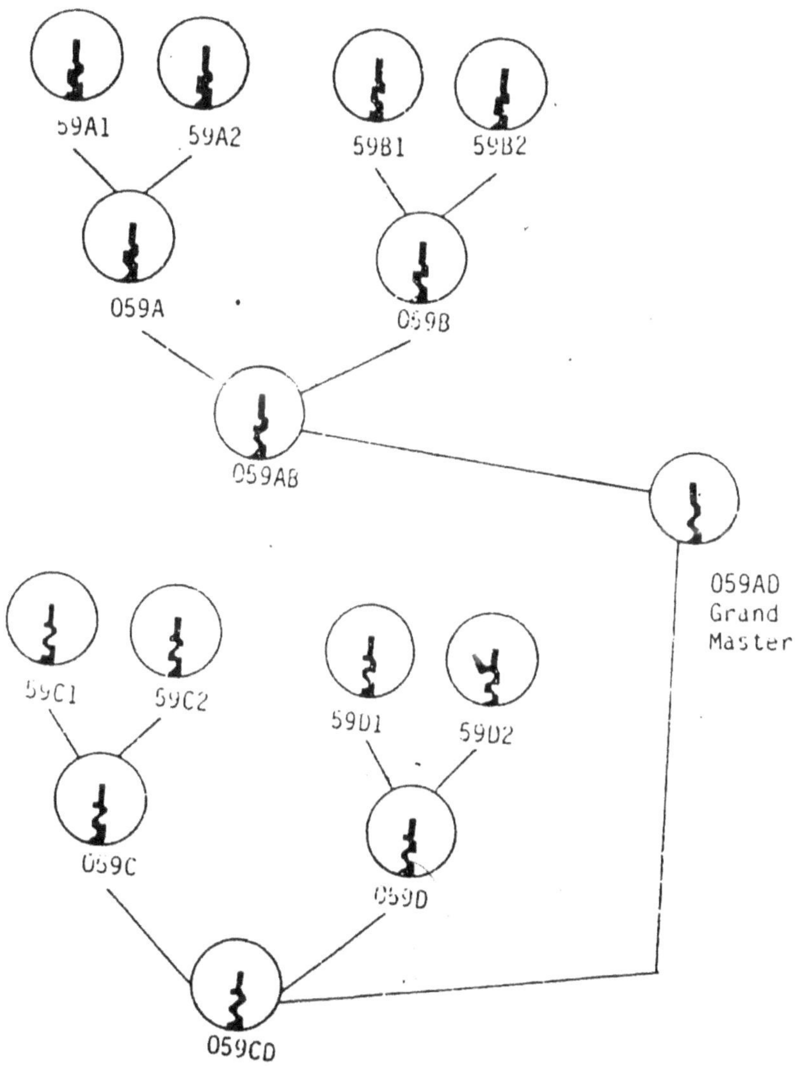


CORBIN 59 SERIES

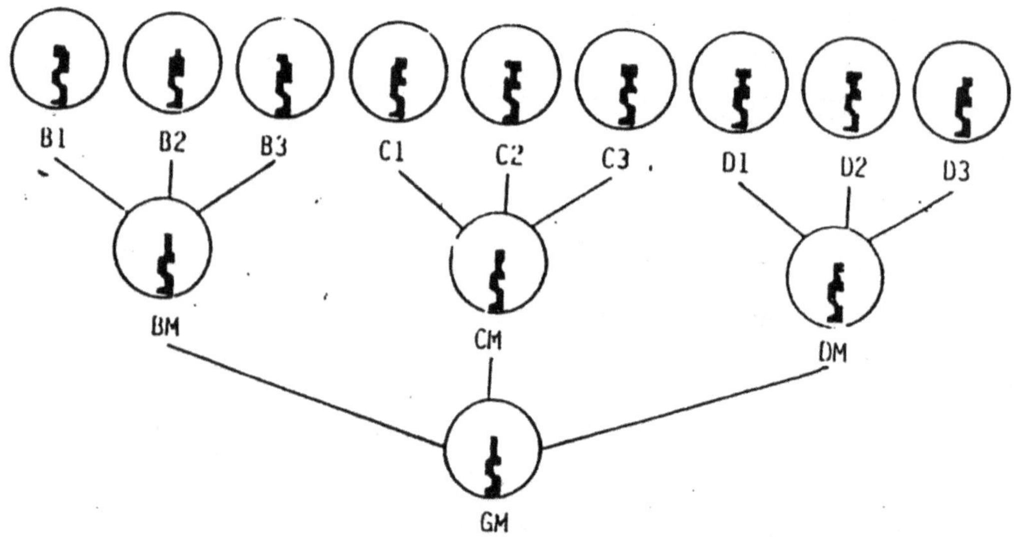


DEXTER
67 SERIES (5 PIN)
62 SERIES (6 PIN)

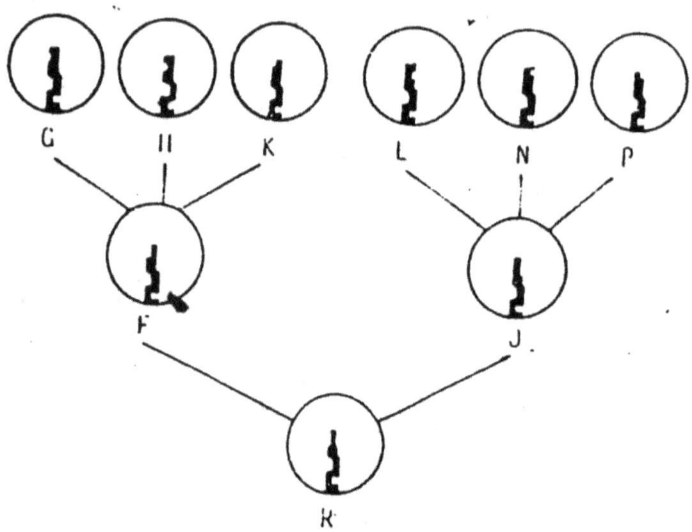


Falcon/Weiser (REGULAR)

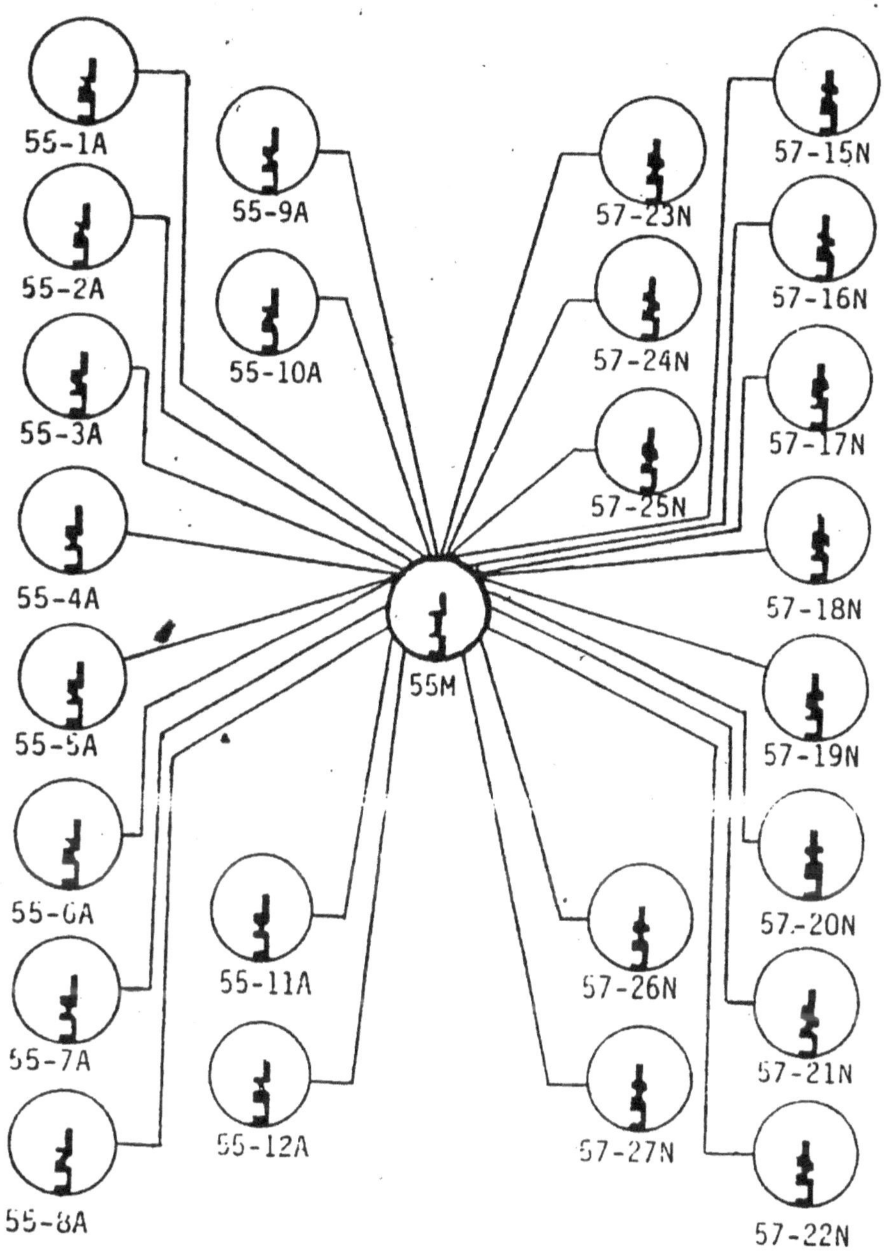


RUSSWIN
752 SERIES
6960 SERIES
(Taylor #'s Shown)

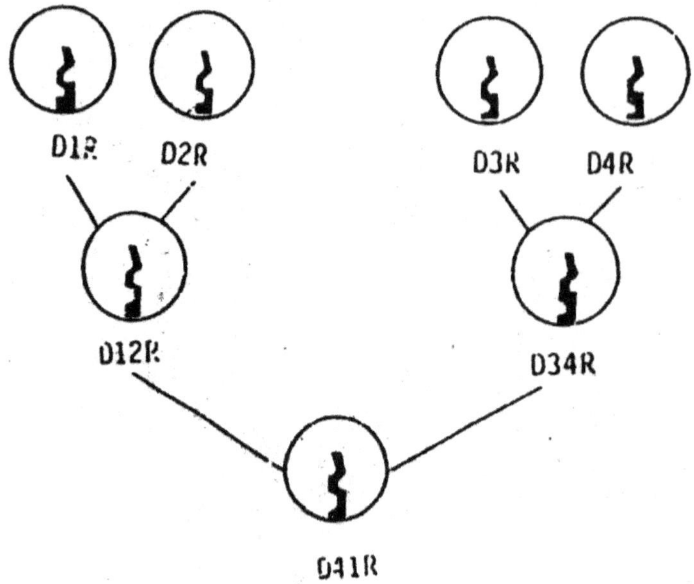


RUSSWIN

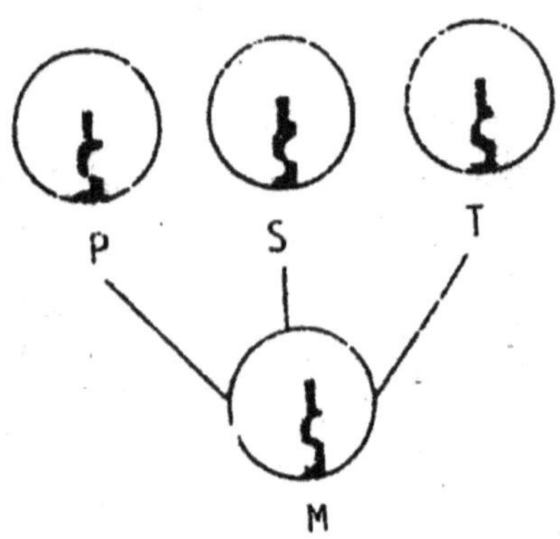


RUSSWIN

If you look closely at any of the keyways for the primary keys (ones the core or cylinder is actually broached to accept properly, not the master blanks' keyways) you can usually see the pattern of how the keys work.

On Corbin 59 series, there are two grooves on one side and on groove on the other. These grooves either go "up" or "down" while the master keyway is cut round to bypass both.

If the top groove on the side with two grooves goes "up" it is either C or D, and if it goes down it is either A or B. If the bottom groove on that side goes up it is a 2, and if it goes down it is a 1. The single groove in the middle of the key on the other side differentiates between C and D or A and B. If it goes up it is a B or D, and if it goes down it is an A or C.

Therefore, if we have a key where the two on one side both go down that indicates either A1 or B1, and if the groove in the middle on the other side goes up it determines that it is a 59B1 keyway.

By the way the top master keyway, 059AD also passes the 60 keyway, which uses a V cut in each of the three positions, which is passed by the round cut in all three positions. It does NOT pass the 70 keyway, because it is the REVERSE of the 60, and therefore has the cut positions on the wrong side of the key.

On Sargent double lettered series primary keyways, LA-LC, LE-LG, and LL-LL, RA-RL, etc., if the bottom groove is "down" and on the left side of the blank (thumb side if holding the blank in your right hand), the keyway is LA, LB or LC. If the bottom groove is "up" and on the left, it is LE, LF or LG and if it is a V shape, it is LJ, LK, or LL.

If on the other hand it is on the right and "down" the second letter possibilities remain the same but it is an RA, etc.

The three positions at the top portion of the keyway determine the exact letter between the three possibilities. There are three places for a small square groove to be placed: Top, Middle and Bottom. Top position indicates A, E, or J. The Middle position indicates B, F, or K. The Lower position indicates C, G, or L.

Understand that the "master keyways" are key blanks only, NOT cylinder keyways. Cylinders are typically only available broached in the primary keyways. Schlage is an exception to this rule. They have an H keyway available that the key passes the C, CE and E keyways, for example.